THE RATIONAL FACTORY

Studies in Industry and Society

PHILIP B. SCRANTON, SERIES EDITOR

Published with the assistance of the Hagley Museum and Library

The
Rational Factory

Architecture, Technology, and Work in America's Age of Mass Production

LINDY BIGGS

THE JOHNS HOPKINS UNIVERSITY PRESS
BALTIMORE AND LONDON

This book has been brought to publication with the generous assistance of Auburn University.

© 1996 The Johns Hopkins University Press
All rights reserved. Published 1996
Printed in the United States of America on acid-free paper

05 04 03 02 01 00 99 98 97 96   5 4 3 2 1

The Johns Hopkins University Press
2715 North Charles Street
Baltimore, Maryland 21218-4319
The Johns Hopkins Press Ltd., London

LIBRARY OF CONGRESS CATALOGING-IN-PUBLICATION DATA

Biggs, Lindy.
    The rational factory : architecture, technology, and work in
America's age of mass production / Lindy Biggs.
        p.   cm. — (Studies in industry and society ; 11)
    Includes bibliographical references and index.
    ISBN 0-8018-5261-7 (hc : alk. paper)
    1. Plant layout.   2. Mass production—United States.
3. Production engineering.   4. Industrial efficiency—United States.
I. Title.   II. Series.
TS178.B54   1996
658.2'3—dc20   96-10947

A catalog record for this book is available from the British Library.

To Steve

Contents

Preface and Acknowledgments

THIS BOOK GREW OUT OF AN INTEREST IN INDUSTRIALIZATION, common architecture, and the transformation of work in industrial America. As a graduate student in New England, I was intrigued with nineteenth-century textile mills and the other industrial buildings that one finds tucked away in both rural and urban corners of that region. Having spent my first twenty-five years in the Midwest, I was more familiar with the modern, sprawling automobile plant than the narrow, multiple-story factory building of the nineteenth century. As I studied the history of technology and industrialization, I wondered how much factory buildings—artifacts of industrial development—could tell us about the changes that had transformed American industry. What could I learn by studying how and why the builders of factories had shifted to such a radically different architectural paradigm at the turn of the century?

I learned that, at least since the late eighteenth century, the men who owned and built factories had a vision of how factories might work, an ideal of a factory that could work like a machine. The history of industry has been, to some extent, the search for that ideal and has led to the constant rethinking of factory design—its physical shell as well as its internal organization.

The dissertation that lies behind this book was supervised by Merritt Roe Smith and Robert Fogelson at the Massachusetts Institute of Technology. Fogelson first taught this converted social scientist to think like a historian. Smith introduced me to the history of technology and provided support and encouragement without which the project could not have been completed. Other faculty at MIT inspired this work, and I hope that they may see some of their influence in these pages: the most important are Leo Marx, whose work serves as a constant model to me, and David Noble, who showed me the importance of asking hard and interesting questions. Other faculty and visiting scholars helped to create a stimulating intellectual environment, especially Gary Hack, John Staudenmaier, Michael Smith, P. Thomas Carroll, Sarah Deutsch, and Louise Dunlap.

My fellow graduate students did what all good graduate student colleagues do, they cajoled, criticized, and stood by through good times and bad. They included Colleen Dunlavy, Gary Herrigel, George Hoberg, Cindy Horan, Paul Josephson, Sarah Kuhn, Frank Laird, Pamela Laird, Anno Saxenian, and Dick Sclove.

The members of the Society for the History of Technology welcomed me as a young graduate student and supported my work as it matured; they are an amazing group of people, and as a group they have my sincere thanks. The faculty and students of the University of Pennsylvania's History and Sociology of Science Department, where I made long-lasting friends in one short semester as a visiting assistant professor, have continued to be supportive and encouraging. My colleagues at Auburn University have likewise provided continued moral support.

A number of people read all or part of this manuscript in its countless drafts. I am indebted to Robert Asher, Nina Lerman, Carol Harrison, John Staudenmaier, Larry Gerber, Ruth Crocker, Julie Johnson, Jim Hansen, and Steve Reber for insightful comments.

The archivists and staff at the Massachusetts Institute of Technology Library, the Harvard University Library, the Baker Library of the Harvard Business School, the Ford Motor Company Archives at the Edison Institute and Greenfield Village, the Ford Motor Company Industrial Archives, the Reuther Library, and the Hagley Library helped in innumerable ways during the course of this research. I also thank student assistants Jessica Dixon and Susan Fraser for the hours in front of the microfilm reader and xerox machine. At different stages of this project,

I received financial support from the Indiana Historical Society, the Kevin Lynch Thesis Award, Auburn University Grants-in-Aid, the Auburn University Humanities Endowment, and the Hagley Museum and Library.

Finally, and most importantly, I thank Steve Knowlton for reading drafts, offering his inimitable critiques, and all the things that have nothing whatsoever to do with industrial history.

Introduction

*The factory should be considered the master tool with which
the factory manager is equipped. If it is not properly designed
and constructed for its work the whole manufacturing function will suffer.*

PAUL ATKINS, *FACTORY MANAGEMENT*, 1926

FACTORY BUILDINGS OF THE NINETEENTH AND EARLY TWEN-
tieth centuries stand in the American landscape as symbols of industry,
projecting powerful images of the enterprises they once housed. The
multistory brick structures of the nineteenth century present a stark contrast to the later generation of sprawling, single-story, concrete and steel
suburban factories that epitomize modern mass-production industries.
The differences in these two conceptions of factory architecture speak
volumes about invisible and often unarticulated decisions concerning
work and production processes. The architectural changes represent an
important step in American industry as industrialists embraced the ideal
of mass production and employed engineers and architects to create a
new kind of factory.

Even a cursory comparison of a nineteenth-century textile mill and a
1920s automobile factory reveals the dramatic shift in thinking. What
happened to cause such extensive changes in the way engineers and
industrialists thought about how to build factories and organize production? Who was responsible for the new ideas? What role did the new
factories play in managerial and technological changes? How were new
ideas incorporated into the actual design and layout of new factories?

The answers to these and other questions add to our understanding of the transformation of American industry.

The lines that open this chapter were written by industrial engineer Paul Atkins in 1926, during a time of rapidly changing ideas about factory production.[1] By that year industrial engineers understood that careful attention to the design of factory buildings could significantly improve productivity. The design of the building had become so important to them that, as factory engineers, they considered it almost as important as interchangeable parts and special-purpose machines, the nineteenth-century developments that made American industry so productive.

By 1926 the nature of the factory building had changed. In the late nineteenth and early twentieth centuries, owners and engineers had begun to build a new kind of factory, and in so doing they recast the idea of what a factory should be. The building had grown larger and more expensive, requiring industrialists to regard it as a major part of the investment in a new industrial enterprise. But it was more than an expensive part of industry: it became a central feature in the planning of large production operations. No longer a passive shell simply to house machines, tools, and workers, the new factory embraced a more complex vision: it became "the master machine," organizing and controlling work. The new factory, as the engineers envisioned it, became part of production technology, helping to solve problems that stood in the way of efficient mass production.

The key to understanding the architectural transformation of the factory in the United States lies in the examination of the nineteenth- and twentieth-century engineering effort to create a rational factory—one that could run automatically as though it were a grand machine. In the rational factory every element of production, including the workers, had to function with precision and predictability. Owners and engineers developed the rational factory in two stages: first they mechanized distinct, individual operations as a way of increasing productivity or reducing skill requirements; then they standardized and regulated the entire production process from the moment when raw materials entered the factory until the finished product left. They accomplished those ends by redesigning the factory and introducing mechanized materials handling.

Some owners and builders of early factories realized that commercial success depended on their ability to understand the factory as a whole rather than as a collection of individual machines and workers. They approached this view through a powerful metaphor—the machine—which, as early as the eighteenth century, engineers employed to guide their work in building and organizing manufacturing. The idea of machine works to guide industrial production seems to us today almost absurdly redundant, but at a time when goods were handcrafted by artisanal labor, it was revolutionary. Though difficult to achieve until the beginning of the twentieth century, the vision of a machinelike factory inspired engineers for more than a century as they constantly sought a better way to organize production for the economic use of labor and materials and supervisory control of workers. By the end of the nineteenth century, engineers realized that, by rethinking the layout of buildings and the way things moved through them, they could approximate a factory that ran like a machine.

Behind the rational factory in the United States lay an intellectual tradition—the Enlightenment's mechanistic philosophy—and a practical motivation—to manufacture goods cheaply. The practical problems confronting the factory manager were largely those of organizing industrial production: moving raw materials and parts into and around the factory, shaping or assembling those materials and parts, and, prominent in early engineering rhetoric, the "labor problem." Although not the sole concern of rationalizers, the labor problem, in its different incarnations, lies at the heart of attempts to rationalize production.

Throughout the nineteenth and twentieth centuries, American industrialists and engineers believed that, if they could eradicate the labor problem, however they chose to define it, their troubles would be over. The labor problem, or the "labor question," persisted as a major concern, but its meaning changed over time. At the end of the eighteenth century and the beginning of the nineteenth, it was perceived as a shortage of experienced artisanal labor. The skilled labor force was too small to do the work of a growing country.[2] Later, during the mid-nineteenth century, the labor problem was defined as a work force lacking "industry," one unwilling to work hard. This definition was combined with, or sometimes replaced by, the concern over unionization, which was sometimes referred to as "the labor problem." Waves of European immigrants, sometimes more ready to organize than their American counterparts, also entered into the definition by midcentury. By the end of the nineteenth century, the term had taken on a slightly different cast as a new breed of managers complained about undisciplined workers, that is, those who would not work according to a strict set of movements and directives as prescribed by the new schools of managerial thought. For late-nineteenth- and early-twentieth-century engineers, the labor question became how to get workers to work the way engineers wanted them to, how to get them to work along with machines, and ultimately, how to get them to work *like* machines.

In 1776, Adam Smith in his *Wealth of Nations* described the economy as a mechanism. Moreover, he introduced the British industrialist to the image of the worker and factory as a machine. In his famous description of the division of labor, he proposed that workers should perform their jobs with the efficiency of a machine: that "reducing every man's business to some one simple operation, and . . . making this operation the sole employment of his life, necessarily increases very much the dexterity of the workman."[3] But Smith's description of the division of labor and his suggestion for standardization of work was not the first. More than twenty years earlier, French engineers had talked about using the "culture of science," by which they meant standardization, codification, and rationalization, to improve industrial work.[4] In 1753, for example, the Royal Porcelain Manufactury at Sèvres was reorganized to increase the division of labor: craftsmen who understood the entire process were replaced by specialists trained in a single step of porcelain making. Thus the mechanistic philosophy of the Enlightenment helped to form the basis of an industrial mentality.[5]

The machine became industry's most prominent metaphor for the next two centuries. In his 1835 *Philosophy of Manufactures,* Andrew Ure, a British writer and scientist, used it as he articulated the premises of what would be the century-long development of the rational factory: "the principle behind the factory system is to substitute mechanical science for hand skill."[6] Like nineteenth- and twentieth-century engineers, Ure recognized that the major difficulty in organizing the factory was not merely a technological one but one of the "distribution of the different members of the apparatus into one cooperative body" and most importantly of "training human beings to renounce their desultory habits of work, and to identify themselves with the unvarying regularity of the complex automaton." Ure anticipated Henry Ford's attitude toward workers. "The modern manufacturer . . . will employ no man who has learned his craft by regular apprenticeship," for that man would be "self-willed and intractable . . . [and] the less fit component of a mechanical system."[7] Ure is saying quite clearly that the worker should become a part of the mechanical system.

Although Ure thought he was describing the situation he saw in early-nineteenth-century textile mills, he was, in fact, describing an ideal far beyond anything that existed at the time. It was an ideal that factory engineers sought throughout the nineteenth century and achieved only in the twentieth. Indeed, Ure foresaw the changes that would occur over the course of the century.

Ure's contemporary Charles Babbage also wrote about the organization of work in *On the Economy of Machinery and Manufactures,* in which he urged the manufacturer to "carefully arrange the whole system of his factory." Other engineers, not as well known as Babbage, expressed similar attitudes toward work and workers. In 1789 another Englishman wrote that "each person is but an engine in the great mechanism."[8] In 1823 still another complained about the unreliability of workers: "stone, wood, and iron are wrought and put together by mechanical methods; but the greatest work is to keep right the animal part of the machinery."[9]

As they created a new society, eighteenth-century Americans imported mechanistic philosophy and other ideas of the European Enlightenment. Rationalism imposed the metaphor of the machine on things social, cultural, and religious: in Europe, God became the clockmaker; the U.S. Constitution was to be "the machine that would go of itself."[10] By the end of the eighteenth century, at least one man actively applied those ideas to American industry. In 1791, Oliver Evans built a flour mill that worked nearly automatically, almost without workers. Thinking about work as a system or mechanism was revolutionary in predominantly agricultural America at the end of the eighteenth century. Preindustrial work was craftsmanship; it was unique, individual work; one worker did not have to depend on the next in order to complete a job. The conception of machinelike work would change that.

The machine became a powerful symbol in the early United States; machines transformed the economy, and as Tench Coxe, assistant secretary of the Treasury, said in 1787, machine technology would become the center of the nation's power.[11] For mechanics and engineers the machine was an ideal to strive for. A machine was predictable and perfectible; it was controllable, nonidiosyncratic, easy to routinize and systematize. This was the control they sought as they built factories and as they rationalized work.

The word *rational* has more than one meaning, but for industrial history the meanings are relatively consistent. Rationalization has been described as systemizing, as centralization, as deliberate and careful planning toward a specific goal, as planning that will allow a business to function in a predictable and secure environment.[12] The rational factory does not simply suggest the use of reason or common sense, nor does it imply the absence of the irrational. It is more than the presence of a rationale. *Rationalization* in industry has a more specific meaning: it refers to the introduction of predictability and order—machinelike order—that eliminates all questions of how work is to be done, who will do it, and when it will be done. The rational factory, then, is a factory that runs like a machine.

My use of *rational* and of the machine metaphor that lies at the center of my definition of the rational factory derives from the Enlightenment's mechanistic philosophy: decisions or activities based on reason, standardized, scientific, predictable, machinelike. *Rational factory* is my term to describe the engineers' vision of a factory that could run like a great machine. I have found no specific reference to a "rational factory" in my research. What I have found, however, leaves me with no doubt about the validity of the term, and I believe that it is used here in a way true to the industrialists' philosophy. Some of the best examples of the early-twentieth-century rational factory are the buildings of the Ford Motor Company, a company that experimented with design and layout more deliberately than most. Henry Ford was more willing than many industrialists to abandon an old building and build a new one either when production volume outgrew the original structure or when engineers' thinking about production advanced beyond the building's capabilities. Consequently he changed factories more often than anyone else during the dynamic period between 1904 and 1920. The Ford Motor Company provides us with a unique record of the steps in the process of building the rational factory in the early twentieth century. The buildings themselves illustrate the stages in the development of mass production and the concurrent changes in the role of the factory building.

The Rational Factory is an examination of an idea as it developed from the end of the eighteenth century to the early twentieth century and of the engineers who used the idea as they built new factories. The idea of the rational factory guided development in large-scale industries whose success depended on control of production. The book is not, however, a history of engineers and their work but, rather, an exploration of the quest for the perfect factory and what it meant for the people who worked within the system as well as those who designed it. This is the story of the recasting of the workplace in the image of the machine that could manufacture goods without regard to recalcitrant workers. Like a machine, the rational factory could be set and adjusted; it could be turned on and expected to work at a predetermined pace in a predetermined manner. This new factory gave workers little choice but to do their jobs in a systematic way, giving up their traditional work habits that so annoyed managers. Its story is a new chapter in the industrial history of the United States.

Rationalizing Production in Nineteenth-Century America

At the end of the eighteenth century and throughout the nineteenth, American engineers, managers, and factory owners strove to make production more efficient by experimenting with new ways of doing work and new ways of organizing the workplace. The motives behind the changes varied—the need to lower costs in an increasingly competitive environment, the desire to gain greater control over the workplace, or the demand for increased precision. During the first half of the nineteenth century, most manufacturers concentrated their efforts on the mechanization of individual operations. During the second half, important innovations were made in materials handling, throughput control,[1] and factory architecture, changes that led the way to the twentieth-century rational factory.

The story of the rational factory in the United States begins with Oliver Evans, the Delaware inventor who built an automatic flour mill in the 1780s (patented in 1790). At a time when the young country was caught up in debate over its economic future, Evans's mill stands out as a harbinger of things to come. The mill on the banks of Red Clay Creek was a marvel of automation; Evans replaced workers with machines and introduced a level of control over production not achieved in other industries for almost a century.

Evans understood and employed one of the important principles that would become a defining trait of mass production a century later—that the movement of materials (grain in Evans's case) through the mill or factory is one of the most important ways to control the speed of production. Evans designed his revolutionary handling mechanisms because he was able to see the mill as a whole, a perception that revolutionized milling and made a conceptual contribution to industrialization almost as important as the division of labor and the American system of manufactures. It provided a cornerstone for the ideological foundation of the modern factory in the United States; Evans's vision of the factory as a whole, as a machine, was rediscovered by industrial engineers one hundred years later and became the guiding principle of the twentieth-century mass-production factory.

Oliver Evans was a natural inventor and a brilliant mechanician in an age of scientific and technological curiosity. His peers included some of the country's leading scientists and inventors: Benjamin Franklin, David Rittenhouse, Thomas Jefferson, and others. Like the work of those other inventors, Evans's ideas were not fully appreciated at the time. In fact, as Hezekiah Niles, editor of *Niles' Weekly Register,* reported, some considered Evans to be the kind of man who "would never be worth anything, because he was always spending his time on some contrivance or another." Venting his frustration at his lack of recognition, Evans himself wrote, "It is not probable that [a man's] contemporaries will pay any attention to him, especially those of his relations, friends and intimates; therefore improvements progress so slowly."[2] Though his contemporaries may not have fully appreciated Evans's achievements, history gives him his due; he is remembered for his automatic flour mill as well as his high-pressure steam engine.

As far as we know, Oliver Evans's career as an inventor began around the age of twenty-two when, after several years as a wheelwright and wagonmaker's apprentice and a short stint in the Delaware militia, he turned to the business of making cards for combing wool and cotton fibers for the textile industry. Evans designed a machine that could bend the fine carding wires and cut them at the rate of a thousand per minute. Though apparently successful at that enterprise, he turned away from it after only a few years and joined one of his brothers in the operation of a general store. During that time the young inventor began thinking about the automatic flour mill. At the age of twenty-seven,

Evans joined two other brothers in a milling business and perfected his plans.[3]

Flour milling was one of the country's principal industries until after the Civil War. The job of milling grain into flour at the end of the eighteenth century was a fairly simple one: The grain was transferred from the wagon or ship to the mill. The miller carried the sacks of grain up one story, where he emptied them into a tub, which was then hoisted by a jack to the top of the mill. The jack required one man below and another above. The grain and meal were moved by hand six more times—to the rolling screen, to the millstone hopper, then to the meal loft, where the wet meal was raked to dry and cool, next to the bolting hopper, and finally to the place where the "richest and poorest" parts of the flour would be mixed together and bagged.[4] Although the milling of grain was easy and straightforward, the process required what, to Evans, seemed unnecessary labor, and he embarked on a campaign to convince the often conservative millers that his inventions would save them money.

For his automatic mill, Evans invented five new machines: an elevator, a conveyor, a hopper-boy, a drill, and a descender. As he described them in *The Young Mill-Wright and Miller's Guide,* they "perform[ed] every necessary movement of the grain and meal from one part of the mill to another, or from one machine to another, through all the various operations from the time the grain is emptied from the wagoner's bag." The new machines carried the grain or meal through the entire process "without the aid of manual labour, except to set the different machines in motion. This lessens the labour and expense of attendance of the flour mill, fully one half." The typical mill employed one hand for every ten barrels ground daily; with Evans's improvements a miller needed only one man for every twenty barrels. "A mill that made forty barrels a day required four men and a boy; two men are now sufficient."[5]

In addition to reducing labor, Evans argued, his automatic mill also did the work better. He claimed that "the meal is better prepared" and that his machines did "the work to greater perfection." He also promised greater efficiency: the machines "save much meal from being wasted" and "they afford more room than they take up." They were also economical in the long run, for they "last a long time with but little expense of repair, because their motions are slow and easy." In addition, "they hoist the grain and meal with less power, and disturb the motion of the mill less than the old way."[6]

Evans emphasized the labor-saving value of his mill for good reason. In preindustrial America securing labor was often difficult, and the general discussion among promoters of industry emphasized the need to replace human labor with animal or machine power. As a consequence, the design of more efficient machines and production methods became a consuming interest for all manufacturers.[7]

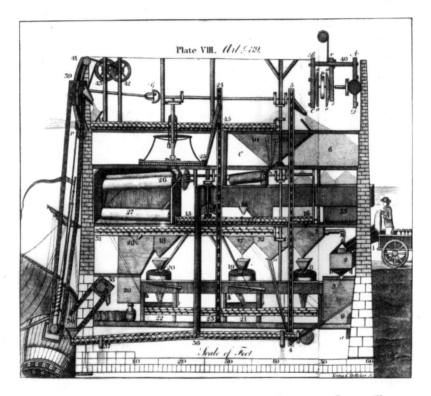

Cutaway drawing of Oliver Evans's early-nineteenth-century flour mill, perhaps the first example of industrial automation in the United States. Grain loaded at one end of the building proceeded through the stones and funnels of the mill to a vessel tied alongside it—supposedly without the need of interior workmen. From Oliver Evans, *The Young Mill-Wright and Miller's Guide* (1795; reprint, Philadelphia: Carey & Lea, 1832). Courtesy of Hagley Museum and Library, Wilmington, Del.

To Evans the obvious way to deal with the shortage and expense of labor was to change the system so that it required fewer workers. By introducing machinery that controlled the flow and speed of work and resulted in almost total predictability of the operation, he rationalized the milling process. The movement of the grain around the mill no longer depended on the speed and brawn of the miller or his helper: a machine now carried it at the same speed every time. The miller could predict exactly how long it would take to grind a given amount of grain; he was no longer dependent on the cooperation of his workers. Thus Oliver Evans's most significant contribution was not in the profits his invention produced for flour millers but in the idea of the automated factory—one that could run as though it were a machine itself.[8]

Views on Technology, Industry, and Labor

Despite the success of Evans's mill, as industry developed in the United States, factory owners were slow to adopt the principle of the rational factory. Evans conducted his experiments in invention and milling in an environment full of question about the future of manufacturing in the United States, and that ambivalence must have contributed to the slow pace of rationalization. Agriculture held a special place in the new country, and many Americans regarded it as "a source of cultural value and a sign of virtue, a moral as well as economic condition."[9] Many were unsure of the role technology and manufacturing would play in a republican society and wondered if it would degrade the yeoman culture as it had Britain's peasantry.

Industry and its accompanying technology became important topics during the Revolution as colonists sought economic independence from Britain. When the rebellious colonists refused to buy England's manufactured goods, they had to produce their own, usually in their homes. After the war some people argued that the new country should develop the industries begun during the Revolution; others said that the United States would be better without the corrupting industry of England.

The agrarians, whose most famous spokesman was Thomas Jefferson, believed that the continued physical and moral health of the country depended on its remaining an agricultural economy and that the young country should "let our workshops remain in Europe." The opposition— represented by Alexander Hamilton (secretary of the Treasury), his assistant, Tench Coxe, and other prominent statesmen—believed that the future of the democracy lay in the development of manufacturing. As we know, the promoters of industry won the debate, eventually persuading even Jefferson, who in 1817 wrote, "I was once a doubter . . . but the inventions of the latter times, by labor-saving machines, do as much now for the manufacture[r] as the earth for the cultivator."[10]

The men promoting industry in the late eighteenth and early nineteenth centuries understood well the significance of machines in manufacturing. In 1789, when Tench Coxe addressed "an assembly of the friends of American manufacturing, convened for the purpose of establishing a Society for the Encouragement of Manufactures and the Useful Arts," he argued that factories run by water, wind, or horsepower and equipped with ingenious machines could operate with fewer hands. Machines thus powered made metals, gunpowder, paper, boards, cloth, and other materials. In addition, they allowed women and children to replace the work of hundreds of male workers in the textile industry.[11]

His concentration on the potential of the machine helped Coxe to address the scarcity and high cost of labor. In his 1794 book, *A View of the United States,* he explained that manufacturers suffered from "the high rate of labour, [and] the want of a sufficient number of hands" and that machines offered a solution. "Machines will give us immense assistance. Combination of machines with fire and water have already accomplished much more than was formerly expected from them by the most visionary enthusiast on the subject."[12]

Also important to the discussion about the availability, cost, and skill of labor was the question of the morals of an industrial population. Jefferson's strongest criticism of manufacturing was a moral one: he argued that America's virtue depended on "direct contact with nature" and that agriculture constituted the only true form of wealth.[13] Although Jefferson and others opposed factories, few opposed technology itself. Even Jefferson was a champion of labor-saving technology on the farm and in the home. He did not realize the potential for change inherent in the use of his nonindustrial machines.

Contrary to Jefferson's position, others suggested that because of the country's excellent farmland, "the farmer and his family have no motive to industry, and idleness is the parent of vice." This unidentified writer concluded in 1817 that gambling and intoxication would spread unchecked unless responsible citizens did something; he suggested that the building of manufactures could provide "industry, good order, and riches . . . and the whole face of the country put on the appearance of a garden." The writer expressed the sentiment of industrial advocates who railed against idleness for the rest of the nineteenth century. Carroll D. Wright, sixty years later, expressed a similar opinion in the 1880 Census of Manufactures: the factory system was "a moral force in the actual progress of civilization . . . an active element in the upbuilding of the characters of peoples," and the factory workers "are kept from falling into habits of idleness."[14] The idleness argument creates a curious juxtaposition to the labor shortage argument. Could there be a true labor shortage alongside an idle population?

Tench Coxe, like others, responded directly to Jefferson's fears with justifications for introducing factories. "Children too young for labour could be kept from idleness and rambling and of course from early temptation to vice by placing them in manufactories."[15] Coxe means here that girls and boys, those too young to begin learning a skilled trade, could be kept productively occupied in an industry where machines required little skill but significant discipline. As we shall see, this notion of machine-discipline recurs over and over during the development of mass production. Throughout the nineteenth century factories had a strict set of work rules regarding tardiness, talking on the job, using liquor and tobacco, fairness to those supervised, and other activities in the workplace. In midcentury one company maintained that the strict moral conduct required by these rules should "convince the enemies of domestic manufacture that such establishments are not 'sinks of vice and immorality,' but on the contrary, nurseries of morality, industry, and intelligence."[16]

More persuasive even than the moral argument was the one concerning political economy. Tench Coxe's promotion of machines was part of the debate about industrialization in the United States at the turn of the century. The debate centered on the future of the economy and culture, whether the new country would remain an agriculture-based economy or build an industrial base. Disruptions in trade with Europe during the War of 1812 strengthened the case for domestic manufactures. Industry's proponents argued that "a nation can be independent only in proportion that she possesses and makes use of the means of producing those things within herself which are essential to the subsistence of her people, and the protection of the state." E. I. duPont supported that position in 1817, as president of the Society of Delaware for the Promotion of American

Manufacturing: "The surplus produce of industry . . . is the only sure foundation of [a nation's] independence and wealth" and the United States must "supply our own wants from our own labor, as to need *no more* from foreign nations than they need from us." After watching the transformation of the town of Wilmington, Delaware, from a decaying port and farming community into a robust industrial town, duPont became a vigorous supporter and powerful spokesman for industry.[17]

The industrial economy of the early nineteenth century depended on skilled artisanal labor, and its shortage led to high wages—nearly double those paid in some parts of Britain.[18] Industrial proponents used the labor shortage as one of their central arguments: Where, they asked, would craftsmen be found to produce by hand the goods the young country needed? The only way to secure an adequate supply of goods, they claimed, lay in manufacturing and the use of industrial machinery which would reduce the dependence on high-priced artisanal labor. Thus some believed that the machine would bring economic prosperity and a higher moral standard to the country. Others feared that industry and its machines would result in a Manchester-like human and social degradation. Although Jefferson probably never believed that industry would do more for the country's morals than working on the land, he did change his position to one of support for industry and its technology. His fascination with machines led him to see them eventually as liberators of the human spirit.[19]

Mechanization in Early Factories

The labor problem as it existed for the first American industrialists played a major role in early factory owners' decisions to mechanize and rationalize. Evans's mill had addressed two elements of the labor problem—it reduced the need for human labor and, by virtue of its handling system, controlled the pace of work. After Evans built his mill, progress toward rationalization was slow. Early industries typically introduced machines to perform individual operations or small sets of interrelated ones. The mechanization of a single operation rationalized only that operation, but such mechanization was an essential part of the rational factory because it changed skill requirements.

Not until owners and managers envisioned the mechanization of *all* operations, including materials handling, could a factory or mill be called truly rational. Nevertheless, many industries took important steps toward planning a factory which would lead them steadily toward a resolution of the labor problem. In the early nineteenth century, three industries in particular reflect these changes: textiles, paper, and firearms. They each experienced the labor problem in a different way, but they all sought solutions through mechanization, and their choices set in motion the industrial response to labor that would continue into the twentieth century. These examples are not meant to represent all early-nineteenth-century industries (habits of nineteenth-century companies varied across region and time) but are intended only to illustrate early efforts to rationalize.

Debates over early industry often focused on textiles, for there Americans clearly saw the consequences of hand production competing with machine production.[20] The factory-made English cloth not only cost less, even after transportation across the Atlantic, but generally offered higher quality as well. Here American proponents found the perfect case to demonstrate the merits of factory production: more, better, and cheaper goods. Forced to compete with cheap, foreign-made goods, American mill and factory owners had no choice but to increase productivity and lower production costs in any way possible. Britain's industrial revolution made the path obvious.

Mechanization began in earnest in the American textile industry in 1790 when Samuel Slater joined with his partners, Almy and Brown, to build the first spinning mill in the United States. Slater's mill, in Pawtucket, Rhode Island, depended on the machines Slater built on the basis of his experience in British mills. In the second decade of the nineteenth century, Francis Cabot Lowell propelled textile mechanization forward with the machines he introduced for spinning and weaving in the integrated shops of the Boston Manufacturing Company in Waltham, Massachusetts.[21]

In the 1820s, Lowell and his partners, the Boston Associates, decided to build larger mills and bought land at the site that would become the city of Lowell. As the new mills grew and prospered, the Boston Associates faced a serious labor shortage, which they addressed in two ways: mechanization and the creation of a cheap labor force. Early successes with machines encouraged the owners of textile mills to mechanize; they quickly learned that, in addition to increasing productivity and "saving labor," textile machines also "cheapened labor" by reducing the skill required of workers. By building more complex machines that simplified human operations, the company could hire employees who needed less experience to produce cloth.[22] Having reduced the skill required of workers, owners felt justified in paying lower wages. By employing young women, who worked for lower wages than men, adding machines that increased productivity, and constantly improving the machines to replace skills, mill owners solved the problem of scarce, expensive labor which pressed them in the early years of the nineteenth century.

The mill owners and managers tried to create more than a system to produce cheap cloth; many sought to integrate all components of production—workers and their housing, machinery and mill buildings. When Nathan Appleton said in 1823, "We are building a large machine I hope at Chelmsford," he referred not to a particular, single machine but to the entire works.[23] With the realization that the entire textile mill might ultimately work as smoothly as a machine, owners began their pursuit of the rational factory.

Once they began building textile machines, mill mechanics quickly learned that production-enhancing innovations earned rewards, and these bright, skilled men had little trouble introducing regular changes that increased speed or lowered skill requirements. The introduction of "stop motions," for example, had a particularly significant impact. Stop motions meant that when something went wrong, such as a thread breaking, the machine stopped automatically until the operator fixed the problem. The worker, therefore, did not have to stand over her machine and watch for breakages; consequently she could run more machines than before. Tending more machines increased the workers' responsibility and stress, part of the price they paid for mechanization. Workers called the addition of more machines the "stretch-out," which was often combined with the "speed-up" or increased operating speed of the machines.[24]

The labor shortage also caused mill agents to turn to a then uncommon labor pool—young women from surrounding agricultural communities. As owners of the Massachusetts mills recruited their young, female labor force, they faced another aspect of the labor problem, the solution to which would prove vital to their success. Having decided to recruit young women from farming communities with no manufacturing experience, the owners realized that they would have to instill in the new employees a factory discipline, essentially forcing them to learn a new way of life. In the early years of the textile industry, the New England

Old Woolen Mill (1812) on the grounds of the duPont Powder Works, now the Hagley Museum, Greenville, Del. A typical American textile mill of the early nineteenth century, this solitary structure housed equipment that performed some but not all of the processes of textile production. Not until later did mills produce cloth from raw fiber. Courtesy of Hagley Museum and Library, Wilmington, Del.

companies created a strict system of supervision of a woman's entire day, both inside and outside the mill. Many authors have explained these rules as the reassurance the owners gave to parents that their daughters would be safe, both physically and morally, in the mill town, but the boarding house system served equally to reorient the women to a new kind of daily routine that suited factory discipline. Infractions of any rules, inside or outside the mill, could result in dismissal. The mills thus created a rigid social structure as well as a new system of production. As textile machinists, managers, and owners constantly introduced new machinery and reorganized mills to improve productivity with so-called unskilled labor, they relied on the "moral police" of the social system to create an "industrious, sober, orderly, and moral class of operatives," a description reminiscent of the arguments put forward by American industry's earliest proponents.[25]

As productivity increased in textile mills, owners found that the old mill buildings could not accommodate the growing numbers of machines and workers. Increased production created new challenges to the coordination of activities in the mill, and that coordination became more and more important to the success of the total operation. To address these concerns, owners built new mills.

The growth in production demanded a shift from the relatively small and inconspicuous mills of the early nineteenth century to large industrial complexes that dominated the economic and social life of small towns. A specific style of building characterized the nineteenth-century New England industrial landscape; because of its development in the textile industry, it came to be called the mill building. Early millwrights designed textile mills around the needs of the industry and the limitations of nineteenth-century technology. The needs were simple and straightforward—water power to run the machines, space for the machines and workers, and enough lighting for the operatives to do their work.

The contemporary technologies of power, construction, and lighting limited the possibilities for meeting the demands. Power transmission technology proved to be the most restrictive component of nineteenth-century mill construction. In order to power one or more floors of machines, the millwright had to plan a network of gears and shafts that carried power from the water wheel to each machine. Indeed, the transmission of power usually proved more difficult than its actual generation. The requirements of power transmission and distribution led to the practice of segregating operations by floor and arranging the machines in rows.[26]

Lighting posed additional problems. Early nineteenth-century textile mills relied on natural light and supplemented it with oil lights as necessary. The artificial light, needed in cloudy weather and every morning and afternoon during the short winter days, was a poor substitute for sunlight. Consequently millwrights designed buildings to capture as much natural light as possible. In a wide building, the center section of the shop floor would never receive enough sunlight, so owners kept building long, narrow mills. The average mill building measured 30 feet in width; lengths varied according to production needs.

The thick walls of contemporary construction technology complicated the lighting problem. Wood or masonry buildings were necessarily built with load-bearing walls. The Durfee Mill in Massachusetts, typical for its time, was built in 1875 by mill engineer Frank P. Sheldon. It consisted of a five-story stone building, 376 feet 6 inches long and 72 feet wide. The first-story walls were 2 feet 6 inches to 3 feet thick; the second-story walls were slightly less thick, measuring 2 feet 6 inches to 2 feet 8 inches; upper-story walls continued to narrow until the fifth story, where they were 1 foot 10 inches thick. The size of the windows followed the same pattern: all were 4 feet 5 inches wide, but their height decreased from 8 feet 4 inches on the ground floor to 7 feet 4 inches on the fifth.[27]

The long, narrow, multistory building emerged as the only practical solution for both the light and power requirements.[28] The mill building style accommodated most manufacturing operations of the period and became the standard factory building, merely built large or small according to a company's production volume. Textile mill owners learned quickly that mill design, along with mechanization and work organization, directly affected profits. As other nineteenth-century industries grew, they resolved their spatial and organizational problems in ways both similar to and different from the textile industry.

Papermakers, though working on a smaller scale than New England textile mills, also introduced labor-saving machines to address labor scarcity and the consequent high wages, which combined to increase the cost of paper and impose limits on production. The motivations that pushed papermakers to mechanize differed from those of the textile mill owners, and the consequences for paper workers were significantly different. Machines in the textile industry did more than simply lower costs; they were part of a system meant to create a docile, low-skilled work force. By contrast, historian Judith McGaw suggests, papermakers introduced machines not to lower skill requirements but to raise productivity. Unlike their counterparts in textiles, many paper workers retained their high wages and skilled jobs.[29]

In the 1820s buying and processing rags accounted for the greatest expense in the manufacture of paper; the next greatest was labor. The industry's labor costs increased in the 1820s, and in the early years of that decade, papermakers found it easier to secure capital than to find qualified workers. A company could do nothing about rag costs, but it could and did lower labor costs by introducing machines. The early paper machines in the United States were very expensive, but they eliminated several jobs and in just one year saved the company far more than their initial cost. In 1827 the Laflin brothers installed the country's first paper-making machine in western Massachusetts; over the next two decades, owners of other paper mills followed this turn to mechanization. Paper-makers introduced machines slowly because some processes proved difficult to mechanize and others were not immediately profitable.[30]

The paper industry grew as demand for the new, cheaper, machine-made paper increased. Mechanization and the industry's growth fed on each other. As more companies adopted the new machines, productivity increased, and paper mills built new buildings to accommodate their expansion. The size, weight, and vibration of the new machinery strained the older buildings even more than machines in the textile industry had. Although some of the paper mills' structures followed the basic mill pattern, the papermaking machines, beaters, and washers required single-story buildings with stronger foundations.[31]

As in textile mills, mechanization of papermaking did not result in the level of rationalization that Evans achieved in his flour mill. Whereas Evans had a vision of an automatic factory, owners of textile and paper mills were primarily concerned with ways to lower their manufacturing costs so that they could reduce prices to consumers. They regularly sought those reductions by reducing their labor costs: hiring workers at lower wages or increasing the productivity of their highly paid workers. Both strategies required mechanization, but the mechanization of a single operation rationalized only that operation, assuring that a worker would perform the operation the same way and at essentially the same speed each time. That regularity helped to lower costs in the textile and paper industries, but it had different implications for the American arms industry.

As the industry that nurtured the "American system of manufactures," arms making demands a place in any discussion of the development of rationalization and mass production. The story of mechanization in the arms industry diverges significantly from that in textiles and papermaking. The arms industry did not mechanize simply to increase productivity, to reduce the number of workers, or to lower costs. Rather, it sought something quite revolutionary—precision manufacture of interchangeable parts. The two federal armories at Harpers Ferry, Virginia, and Springfield, Massachusetts, housed the first experiments in interchangeable parts manufacture. The motivation was the production of better and more easily repaired weapons, and mechanization was the key.[32]

To make guns that were truly interchangeable, armorers had to be able to make every gun identical. The best way to assure such uniformity was to systematize the process so that each armorer performed a given task the same way every time. Early efforts to achieve uniformity did not include machines; they employed a system of gauges and jigs that served first as models and then as checks on finished work. Later, machinery was introduced to reduce error and then to reduce the skill required of armorers.[33] In the manufacture of rifles with interchangeable parts, John Hall, one of the creators of the system, reduced skill requirements to the extent that he claimed to have replaced seasoned armorers with young apprentices.[34]

The arms industry provides a fourth model of mechanization and rationalization. Its concerns differed from flour milling, textile manufacture, and papermaking. Gun making was precision work, and the desire for interchangeable parts required that work to be done better. For Evans the movement of the grain around the mill was the most time-consuming task, the most important one to address in order to make operations more efficient. The textile industry developed machines and managerial techniques that lowered skill requirements, forced conformity, and enforced speed of work. The paper industry sought machines to increase productivity but not necessarily at the expense of workers' skill and authority. The arms industry had different problems: it dealt not with heavy, bulky, hard-to-move materials but with relatively light materials requiring precision work. Nor was it dependent on lowering costs and increasing productivity; the aim of rationalization in the armory was consistent precision in the final product.

Despite the different needs in these four industries, managers and engineers employed the machine as a model to help transform the way work was organized and the way jobs were performed. The idea of rationalizing work, of modeling work on the image of the machine, could be applied to almost any job. It pervaded American society to the extent that its influence appeared in so remote a context as the Southern plantation. Adopting the philosophy of the Northern industrialists, Bennet H. Barrow, a Louisiana plantation owner, wrote in his diary (1836–46) that he considered his plantation "a piece of machinery" whose separate operations were controlled by different individuals on the plantation. If everyone exercised discipline, the machine could work smoothly. Southern agricultural journals promoted a "plantation ideal," advocating principles of management much like those of Northern industry; the ideal plantation should be as efficient as the mill or armory. Successful plantation owners realized that the productivity of their plantation depended on the skills and cooperation of slaves.[35] A minority of owners who read the journals and subscribed to their prescriptions systematized their plantations by developing a strict set of rules and disciplinary actions. All owners had rules, but the systemizers hoped that theirs would "reduce everything to a system," thus improving efficiency and accountability. The rules, much like those in the factory, were meant to replace the slaves' independent thinking. The promoters of the rational plantation management system hoped that the rules would provide a more perfect system in which the need for physical punishment would be eliminated. But ultimately the success of the plantation depended less on the systemization of management than on the willingness of the slave to work.[36] Slaves, like factory workers, held ultimate control over productivity, and either their cooperation or control over them was essential. The rational factory sought productivity through control.

The flour milling, textile, paper, and arms industries reflect the major themes in early efforts to rationalize; through mechanization and reorganization the rationalizers sought control of the speed of production and output, a reduction of labor costs, precision and a better product, or control of labor. Industries such as shoes, sewing machines, and furniture introduced machines for the same reasons. In comparing the four early industries, we see the different motivations behind mechanization and rationalization in the early nineteenth century. Oliver Evans clearly sought to gain efficiency and productivity as he introduced machinery to move grain around his flour mill. New England textile mill owners hoped to achieve a factory system that would enable them to produce cloth with as little male artisanal labor as possible. Their new system not only lowered labor costs but also helped to set the pace of work. The owners of paper mills, who would rely on skilled male craftsmen throughout the century, wanted to raise productivity and thereby reduce labor costs proportionally. Armorers, aspiring to manufacture guns with interchangeable parts, had to increase the precision of their work. They first rationalized their operations through gauges and jigs used with hand tools, then later they introduced sophisticated machines.

Rationalization through Materials Handling

The second half of the nineteenth century witnessed major breakthroughs in American industry as new technology and new principles of industrial organization continued to change production styles. Mechanics and a new breed of engineer made changes aimed at increased efficiency, greater output, and higher profits. We can identify two lines of development toward the rational factory during those years. The first—the continued progress of special-purpose machines and the American system of manufactures—changed metalworking industries such as firearms, sewing machines, and agricultural equipment.[37] The focus of this section, however, is on the second development—techniques for processing and the handling of materials—which transformed industries such as canning, meatpacking, and steelmaking. These industries made their innovations independently of one another, according to their own needs. The innovations later formed an important part of the twentieth-century industrial synthesis that became modern mass production.

Processing industries require a different kind of work from mechanical or production industries; they depend on different skills because workers perform fewer and simpler tasks.[38] An industry like arms making needed experienced artisanal workers because the work required a high level of judgment, dexterity, and resourcefulness.[39] The manufacture of guns required close tolerances for the arms to work reliably. Processing industries, on the other hand, consisted of many jobs that required little or no judgment that would affect the final product; the work required careful judgment of only a few workers. Slaughtering and butchering hogs allowed a wide tolerance; pork was cheap, and most of it was ground into sausage, so the cut did not have to be precise. Beef processing required skill in a few select jobs because the hide lost value if torn, and beef was sold as expensive cuts only if the butchering was done well. Similarly, the canning and steel industries required skilled judgment in only a few jobs. Consequently, continuous-process industries advanced toward rationalization faster because the technology they needed did not have to replace the same level of skill and decision making as the machines in the production industries. Once the processing industries developed the necessary technology, speed and volume increased quickly and dramatically. The mechanical industries improved their efficiency, but because of their more complex operations, rationalization was harder to achieve.

In the late nineteenth century, the processing industries took important steps in the development of mass production when they recognized and addressed the significant role of materials handling. Meatpackers,

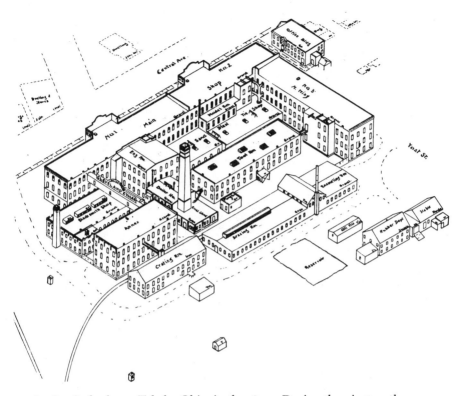

Lozier Cycle shops, Toledo, Ohio, in the 1890s. During the nineteenth century builders and industrialists did not so much rethink the factory as
expand it to suit growing and varying needs. From *Engineering Magazine* 11
(1896): 288; photo courtesy of Hagley Museum and Library, Wilmington,
Del.

for example, understood that next to the few special-purpose machines
that could replace skilled workers, the movement of their raw materials
through the factory would be their most important labor-saving measure. Processing engineers knew that flow offered the key to turning the
processing plant itself into a great, efficient machine.

The concept of throughput is vital to understanding the changes in
this period, changes that serve as the foundations of mass production.
Alfred Chandler argues that throughput, as seen in the refining and
distilling industries, "demonstrates the basic axiom of mass production"
and that by intensifying the speed of materials through processing, a

company gained economies of scale and lowered unit costs.[40] In these industries, increasing the velocity of throughput was more important than adding more machines.

Meatpacking provides one of the best examples of the importance of throughput in late-nineteenth-century industry and illustrates the link between Evans's mill and modern mass production. The industry bears comparison to Evans's mill in the character of the final product—there was nothing to improve. Packing houses could not produce a "better" cut of meat; they could simply hope to do it more cheaply. Consequently, economy and efficiency became essential as each company tried to beat the competition's price in order to survive.[41]

Beginning as an industry composed of small merchants, early-nineteenth-century meatpacking operated as a local business usually separated into its two components, slaughtering and packing, which were sometimes on opposite sides of town. By the 1830s packers realized that time and energy were wasted in moving animal carcasses across town and began to do their own slaughtering, thereby eliminating a large expense. By 1850 nearly all packing houses slaughtered and packed under the same roof.[42]

Like many manufacturers, the packing industry sought increased efficiency through division of labor and, as early as the 1830s, had begun to divide up the work so that twenty men killed, scraped, and gutted each hog. The division of labor was so striking that nearly every visitor to the packing houses noticed it. These measures increased processing speed and accounted for the "human chopping machine" described by Frederick Law Olmsted when he visited Cincinnati's plants in 1857. Olmsted commented not on an inanimate machine but on a set of human hands that could dispatch a hog in thirty-five seconds; it consisted of "a plank table, two men to lift and turn [the table], and two to wield the cleavers."[43]

As part of their cost-cutting efforts in the highly competitive environment, packers, like other industrialists, sought ways to reduce the number of workers in the plant and to increase the productivity of the smaller work force. Handling technology made a significant contribution to that increased productivity: overhead conveyors, endless chains, and "moving benches" eliminated manual handling of carcasses through the packing house. Because other operations, such as killing and cutting, were difficult to mechanize, the cutting line, or "disassembly line," underwent the first changes that contributed to labor savings. The first disassembly line was introduced in the 1850s and, like Evans's system, moved work through the plant, almost eliminating the slow and cumbersome human handling of carcasses. At first powered by hand, the early overhead line worked on the same principle as the later mechanized version. Once conceived, it took only a few years to progress to its automatic conclusion.

The line not only saved time because the workers did not have to handle the animals but also controlled the speed of work, a key to rationalization in the nineteenth and twentieth centuries. In fact, the steam-powered packing house line presaged the automobile assembly line; one packing house representative said in 1903 that the line prevented "the slowest man from regulating the speed of the gang." And, "if you need to turn out a little more," said a Swift supervisor, "you speed up the conveyers a little and the men speed up to keep pace."[44]

Moving a carcass around the plant by hand was difficult and inefficient. In the 1850s, after scraping off the hair, workers hung up a carcass by stretching its hind legs apart using a stick called a gambrel. Three men then carried the carcass away to another part of the building, two of them holding the front part on their crossed hands and the third grasping the gambrel.[45] Later they simply hung the hog by the gambrel on the overhead line, a set of parallel wooden tracks attached to the ceiling, on which the carcass could be pushed along to the next operation. These overhead lines were common in packing houses, especially in Cincinnati, by 1850, and as early as 1860 a traveler even described the industry as "semi-mechanical." Wadsworth, Dyer, and Company of Chicago installed a steam engine in 1850 to aid in handling the live animals and carcasses. A rope attached to the engine pulled a resistant bullock into the killing room and a few minutes later lifted the carcass onto beams to cool.[46]

The early packing houses worked only in the winter. When refrigeration became available and affordable in the 1870s, packers continued operations through the summer months. The ability to work twelve months led to dramatic growth in the industry. After 1875 a handful of Midwestern companies took advantage of railroads and refrigeration and began to centralize the industry. Half a dozen companies came to dominate the trade, with Swift and Company and Armour and Company as the largest.[47]

Despite the impressive "human machine," packers, like other industrialists, sought machinery that would perform specific jobs. As they

began to mechanize, most industries first developed machines that could replace workers' skill and only later thought about mechanizing handling. The packing industry sought mechanical replacements for skilled workers *after* developing a handling system. According to the general superintendent of one packing house, the single most important mechanical improvement in the packing industry came in 1876 with the patent of the first hog-scraping machine, which did what had been an "irksome . . . and tiresome job." The machine held vertical and horizontal scraping reels that revolved as the hog passed through. It was important enough to the industry that it attracted copiers, and new, improved hog scrapers appeared within a few years.[48]

The push for efficiency through technology was so successful that in 1888, Swift boasted that in its plants so much of the work was done by machinery that "a workman rarely has occasion to touch the parts intended for food purposes." In 1900, Armour claimed that "the modern packinghouse eminently exemplifies scientific, commercial methods," and "has logically displaced the smaller slaughterhouses by the application of economies only obtainable in extensive operations."[49]

By the late 1880s the slaughter and butchering of a hog was organized along lines later to become famous in the auto industry. By the beginning of the twentieth century, the work process, the technology, and the plant had become so important that one member of the industry deemed it necessary to write a lengthy treatise: in *The Modern Packing House*, F. W. Wilder suggested the vital role of plant design in the packing process.[50]

In plant design meatpacking once again foreshadowed mass production in the auto industry. Wilder, a former superintendent for Swift, described the important role that plant design played in efficient processing, explaining that the efficient plant took advantage of gravity in planning the flow of operations. The animals should be driven up an inclined platform to the killing rooms, an arrangement that totally eliminated the need to move fresh carcasses. "Gravitation is the cheapest force we have at our disposal, hence it is . . . best to have the cattle killed on

Opposite page: Innovative handling technology in a meatpacking house at the turn of the century. Live hogs, lifted from pens, become carcasses that move through the plant via a gambrel hanging from an overhead track. From F. W. Wilder, *The Modern Packing House* (Chicago: Nickerson & Collins, 1905); photos courtesy of Hagley Museum and Library, Wilmington, Del.

the upper floor and pass the carcass and the offal . . . by gravity to a lower floor."[51]

The packing industry was the first to follow so directly in Evans's footsteps. Meatpackers first introduced innovations in handling because it was such an expensive part of the process, and they mechanized operations only later. Unlike other industries, packing was slower to mechanize its most skilled and expensive jobs. In fact, not until the twentieth century did some of the most highly paid jobs succumb to mechanization. There was no single way to rationalize industrial production or processing.

Like the meatpacking industry, canning prospered after the Civil War. Gail Borden and other canners began building their businesses during the war, and afterward they built empires not only on the rapidly growing demand for canned foods but also on their ability to make canned foods cheaper through the new technologies and new factory organization. Canning companies faced problems different from those of meatpacking and began their rationalization by mechanizing the most highly skilled operations; only later did they move on to questions of handling and throughput.

After the fresh food reached the processing plant, canning consisted of three basic operations: the manufacture of tin cans, the filling and sealing of the cans, and the processing of the filled cans to sterilize the contents and make them safe for consumption.

Can fabrication constituted the greatest expense in the industry, one that had to be passed on to the consumer and made canned foods too expensive for most people to buy. The high cost made can making an attractive process for mechanization, and it was the first part of the operation to be subjected to the changes brought by new technologies.

Before 1870 an expert can maker, working with hand tools only, could make one hundred cans per day. Because the job required significant skill and experience, can makers could demand high wages. Most canneries hired their own can makers to prepare cans for the season, but by the 1860s the first can factories began to appear and with them the seeds of change for the industry. These factories were not mechanized; they simply made a large number of cans by hand. By 1870, however, Baltimore can factories boasted new stamping machines "used to stamp out and form the ends of cans," significantly increasing the production of each worker. With the new machines, an experienced worker could produce five hundred to seven hundred cans per day, and a few of the best could make as many as one thousand per day. A few years later, innovations increased production again, allowing one man to produce an average of fifteen hundred cans per day. These innovations eroded the once privileged position of the expert can maker. The final blow was Edwin Norton's continuous can-making line, introduced in 1883. The machine started by stamping can ends from flat pieces of tin plate; it formed the entire can and then tested it for leaks. Twelve nonexpert, low-paid boys tended the can line and produced thirty thousand cans per day.[52]

By the time Norton introduced his machine, most canneries bought tin cans from factories rather than making them themselves. By 1904 most of the cans in the United States were made by just two companies, American Can Company (incorporated 1899) and Continental Can Company (incorporated 1904). With an abundance of cheap cans, canners began to mechanize processing functions.

By 1887 the Cox capping machine made the job of filling and soldering the caps easier and eliminated the most skilled of the canning workers. Early tin cans did not leave the entire top off before the can was filled but simply left a hole in the top large enough to add the filling. A capper then covered and soldered the can closed. Only a skilled workman could cap quickly and accurately, and a good capper could count on high wages— in the 1880s as much as fourteen dollars for a twelve-hour day. The Cox capping machine eliminated those high-paying jobs, and the cappers in some plants responded with violence: breaking the machines and even burning down entire plants.[53]

The canning industry, looking for further cost-cutting techniques, turned to handling. Sometime before 1885 supply firms began to sell rails on which cans could be easily rolled through the processing sequence. The rails lowered processing costs and significantly reduced the price of canned food. By 1890 the E. F. Kirwan Manufacturing Company advertised a number of important machines for the industry. A corn cob conveyor moved cobs by machine rather than human power. Warfield's can dipping machine automatically filled cans or jars with brine, syrup, or other liquid and employed a chain conveyor to move them. The Burt continuous can wiper used a sprocket chain to move cans under revolving brushes to prepare them for the capper. "The feed is continuous," claimed the catalogue, "and can be adjusted for two or three pound cans, the brushes raised or lowered by the operator."[54]

Machines changed other parts of the industry. Rural inventors around the country experimented with machines to pick and clean crops in the field. Corn cutters removed corn from cobs as early as 1875; the pea podder was introduced in 1886; the pea viner by 1890 shelled the peas while they still hung on the vine; and machines picked the stems and blossoms from berries by 1904.[55]

The nature of the canning industry also allowed its owners to rationalize in a manner similar to the way Evans had in his mill. The canning process consisted of a few relatively simple operations, and once the most highly skilled jobs—making and closing the can—were mechanized, no skilled or well-paid jobs remained. All jobs in the canning industry yielded quickly to mechanization. As in Evans's mill and other mills and factories, after machines could perform or aid in individual operations, the principal route to lowered costs lay in moving raw materials and partially completed operations around the factory.

The greatest of the late-nineteenth-century processing industries—steelmaking—addressed issues of materials handling and plant design more deliberately and aggressively than other contemporary industries. The immense volume and weight of coke, limestone, and ore forced steelmakers to experiment with better ways of moving materials and organizing work. By the early twentieth century, steel mills had succeeded in rationalizing production: with a fraction of their former work force they operated as smoothly and efficiently as a machine. Between 1870 and 1900 steel rose from an inconsequential industry to the country's dominant manufacturing enterprise as a result both of the demands of a rapidly industrializing society that needed more and more steel and of the technological innovations that yielded greater productivity and lowered costs.[56]

As late as 1880 an average furnace produced about fifty tons of steel per day, a volume not easily increased with the technology of the day. Laborers moved the ore, coke, and limestone in wheelbarrows—large steel bins weighing eight hundred pounds empty and twenty-one hundred pounds when loaded—several hundred feet from the storage bins to the furnace, a slow and heavy job and an inefficient one with traditional plant layout. A vertical hoist lifted the materials to the top of the furnace, where skilled top-fillers charged the materials into the furnace by hand. Because charging required careful distribution to ensure standard furnace results, manual charging continued to be the preferred practice until 1895.[57]

The industry grew in the late nineteenth century by increasing the size rather than the number of plants; companies strove for economies of scale which would allow them to compete in the aggressive industry. As steelworks grew larger, manual handling of materials severely limited daily production. By the end of the century, handling of raw materials was the major concern of the industry, followed by integrating production and mechanizing rolling mills.[58] Two examples demonstrate the impact of innovations in plant design and handling on productivity and labor issues: Andrew Carnegie's Edgar Thomson Works (1875), designed by Alexander Holley, and the Sparrows Point complex of the Maryland Steel Company (1891), designed by Frederick Wood.

Alexander Holley, *the* designer of Bessemer works, designed all the Bessemer plants in the United States except for the first in Troy, New York.[59] His early designs generally followed the plan of the Troy works. That changed when he designed the Edgar Thomson works for Andrew Carnegie's Pittsburgh steelworks. Completed in 1875, the Thomson works began to change the way steelmakers thought about the organization of steelmaking. Holley considered the Edgar Thomson works, where he started fresh on a bare piece of land, to be his greatest design. As he later described the design process, he "started from the beginning . . . took a clean piece of paper" and on it "drew the railroad tracks first and then placed the buildings and contents of each building with prime regard to the facile handling of materials so that the whole became a body shaped by its bones and muscle rather than a box into which bones and muscle had to be packed."[60] Carnegie soon recognized the success of Holley's design; in 1881 the works made profits of 130 percent of invested capital.[61]

A few years after designing the Edgar Thomson Steel Works, Holley undertook the task of reorganizing Bessemer steelmaking. All Bessemer plants suffered from limited output because the converter could not be used constantly. One problem lay in the necessity of cooling the converter between batches, the other in the difficulty of moving the finished steel around the plant. Holley increased the number of converters used at the works and also introduced materials handling technology. Until the 1880s, when he introduced power equipment for handling the steel, moving the ingots away from the converters was the major limitation on the speed of the converting process. Steelmakers introduced similar changes in handling in rolling mills. Steam or electric power replaced human muscle for lifting and carrying and before long replaced most of the workers. By the turn of the century, fewer than a dozen workers operated a mill rolling three thousand tons a day.[62]

Sixteen years after the Edgar Thomson works opened, the Sparrows Point works began production in Delaware. Sparrows Point, designed and run by Frederick and Rufus Wood, also demonstrated the importance of organizing handling in order to control throughput. Like Holley, Frederick Wood experimented with plant layout and succeeded in making the Sparrows Point works more compact and centralized than other works. But Wood took mechanization and organization a step farther than Holley; he also concentrated on "mechanizing" his work force, subordinating the workers to the machine as much as possible. Wood went farther than most when he modeled the workday after the continuous workings of his machine and introduced the twenty-four-hour operation with two shifts: the day shift worked eleven hours, the night shift thirteen or fourteen. Shifts changed on Sunday and one crew worked for twenty-four hours.[63]

Wood installed two converters larger than any yet constructed. He could use these only because he had solved problems of materials handling; when designing the plant, he thought about flow of production and set up the works to require a minimum of handling. The same principle applied to the rolling mills, where conveyors kept the steel moving.[64]

Steel itself was a vital component in the design of the works—Wood adapted the steel-frame train shed to be production buildings instead of designing a traditional wood and masonry structure. By doing so he opened up a large, unobstructed work space that could accommodate the cranes that he would use to move materials and finished steel. By the first years of the twentieth century, the steel industry had solved many of its technical problems and boasted almost total mechanization.[65]

The increasingly competitive environment of the late nineteenth century forced most American industries to think harder about mechanization, factory design, and workers. While many mill and factory owners explored ways to rationalize production, those in the processing industries experienced the greatest success. The processing industries moved toward the rational factory faster because they required fewer operations and less skill than the production industries. Because moving materials through the plant is such a large part of any processing activity, once industries such as packing, canning, and steelmaking understood how to achieve continuous movement, they had made significant progress toward rationalization. Advances in handling technology addressed inefficiencies by replacing human muscle with machine power; they also addressed the labor problem by helping to set the pace of work and reducing the size of the work force.

General advances in American industry during the second half of the nineteenth century were summed up by an anonymous writer in *Iron Age:*

> The old system of educating master workmen is passing away, and this
> fact is due to several causes: the tyranny of trades unions, inventions
> in labor saving machinery, improved methods of conducting work,
> inflation of values, and the influence of the late war, during which un-
> skilled labor was employed to do work that was before trusted only to
> experienced men. The system of a division of labor, of men working
> "in teams," as it is called is spreading rapidly. The work is divided up
> as far as possible, so that one man does the same thing all the time,
> and, if he leaves, his place can easily be filled.[66]

An astute observer, the author described what he saw as a sea change in the way work was done. What he could not have known was that the changes he described would continue to develop into a production system that would redefine the nature of industrial work.

By the end of the nineteenth century, industrialists understood that the success of manufacturing lay in mechanization. Special-purpose machines were already helping to build guns, sewing machines, bicycles, and other goods, and handling technologies were revolutionizing the processing industries. In many industries a new kind of engineer also knew that mechanization had to go farther than special-purpose machines, beyond individual operations. As industrialists and engineers sought greater efficiency, they began to understand what Oliver Evans had learned almost one hundred years earlier—that the flow of work determined manufacturing time. The movement of materials around the shop floor could be managed, they found, by mechanization *and* by better design and organization of the factory. By the end of the century, a new group of engineers made these concerns the focus of their work.

Industrial Engineers and Their "Master Machine"

At the Chicago meeting of the Western Efficiency Society on May 26, 1917, a group of engineers launched the Society of Industrial Engineers. The group's purpose was to aid the government in war production and to provide a means "whereby the best minds working together can promote efficiency and industrial management." The first board of directors included men who had already established themselves as professional engineers and others who would shape the new field of industrial engineering in the years to come; they included Charles Going, Frank Gilbreth, Harrington Emerson, Dexter Kimball, Morris Cooke, and Charles Day.[1]

The creation of the new society marked a turning point for the young profession; the men who called themselves industrial engineers finally felt confident that their work, engineering the factory, constituted a legitimate branch of engineering. Many mechanical engineers had started practicing industrial engineering in the late nineteenth century when they began to explore new ways to improve factory production through experiments with management and efficiency. They began to work as general-purpose managers and industrial consultants and prided themselves on their ability to design and organize all operations in the factory. These men applied their engineering perspective and skills to every element of the factory: building design, shop floor layout, machines, materials handling, lighting and ventilation, and management of workers. As they engineered the factory, they redefined it. Once merely buildings in which men and women labored to produce goods, the factories run by industrial engineers became something more. Inspired by the mechanical ideal of the eighteenth-century, these men followed a new vision— the factory as machine. Their success in realizing that vision paved the way for modern mass production.

The Engineers

Industrial engineering developed along with the rapidly growing industries of the late-nineteenth- and early-twentieth-century United States. The mechanization of nineteenth-century industry, the construction of larger factories, increasing division of labor, and greater commercial competition led industrialists to experiment with efficiency techniques. Two of their most troublesome problems were managing labor and organizing the flow of production as it outgrew nineteenth-century practices.

By the third quarter of the nineteenth century, in response to the changing scale of production, American industrialists, managers, and engineers were talking and writing about what modern scholars call systematic management.[2] During the last decades of the century, the complexity of manufacturing began to create significant organizational problems. Continued division of labor and specialization changed the way the factory worked. Division of labor and specialized machines allowed manufacturers to hire workers who had not been trained in the earlier craft traditions; consequently more and more workers knew how to perform only their own small operation. Similarly, supervisors themselves were beginning to have their own duties divided, a process that created managerial specialization.

The factory's growing complexity combined with managerial specialization meant that fewer and fewer people understood how the factory as a whole operated. With work being handled by many people who possessed only partial understanding of the overall operation, factories were drained of efficiency because orders were delayed or even lost, parts were not where they should be, and general confusion reigned. This evident lack of coordination created a situation that cried out for a new profession. The engineering and management literature urged manufacturers to return control to top management through standardization and "system" and thus eliminate confusion, neglect, and error. These ideas would be addressed by the industrial engineer, who sought to understand all facets of production and, in doing so, to organize and operate the factory in the most systematic and predictable way possible.[3]

In the 1870s and early 1880s, engineers participated in the development of an early management system as a way to help eliminate the confusion and inefficiency in factory production. Systematic management was not a unified, self-conscious movement, as scientific management would be; only in hindsight does it resemble a movement. To eliminate confusion and inefficiency prevalent in so many factories, managers promoted cost accounting and production and inventory control, reduced foremen's authority, and introduced incentive wages.[4]

The successes of the young management movement inspired ideas of even greater control over production and provided fertile ground for Frederick W. Taylor and his popular scientific management.[5] An engineer himself, Taylor tackled the fundamental problems faced by the engineer running a factory, especially workers' motivation. Although he tried to address every element of the manufacturing operation, his prescription for dealing with workers became one of the best-known components of scientific management. Taylor's efforts to control the way workers performed their jobs grew out of his own frustrations after watching steelworkers pace their day's work so that they could restrict daily output. The workers viewed pacing as a way to preserve high piece rates; Taylor saw it as unnecessary inefficiency. Taylor's new articulation of the labor problem continued the nineteenth-century discussion of labor. Whereas factory owners had introduced machines to reduce the importance of skilled workers, Taylor brought a new level of control to the shop floor and new level of job division. He suggested that managers develop a "scientific" way for each man to perform his work (time study) and to select and train workmen "scientifically." He further argued that all work should be planned by management rather than workers and that workers should use tools furnished by the company. Tool ownership was a traditional emblem of artisanal skill; requiring employees to use company tools was a step toward eroding that skill. Taylor's program would also remove as much decision making as possible from the worker's job.

In practical terms, Taylor's proposal led to the breakdown of jobs into smaller and smaller operations; with each successive breakdown, the skill and judgment required of the individual diminished. Scientific management also changed the managers' jobs. Managers had previously relied on personal experience as the basis for decisions, but in Taylor's ideal factory, their judgment would also be standardized. With many managerial jobs being standardized, the person who standardized them, whose job was not yet standardized—the industrial engineer—experienced enhanced status.[6]

Taylor emphasized economy and efficiency in the factory and in so doing, reflected the strong engineering interest in the business side of industry, an interest hardly new with Taylor. Themes of economy and efficiency were common essay and lecture topics in the nineteenth and early twentieth centuries. As Henry Towne suggested in his introduction to Taylor's 1911 *Scientific Management*, "the true function of the engineer is, or should be, not only how physical problems may be solved, but also how they may be solved most economically." Engineers agreed that first of all engineering was a business, an attitude that has been described as the ideal of the engineer-entrepreneur—a professional as interested in profits as in engineering excellence.[7] The emphasis on economy in engineering continued to grow in the early twentieth century as scientific management came to dominate the emerging profession of industrial engineering.

The development of the industrial engineering profession was clearly motivated by economics. Throughout the nineteenth century the idea that the engineer must also be a businessman and economist had been promoted by engineering leaders. As early as 1835, George W. Light described mechanical engineering as one of the "business professions." In 1881 a university professor, Thomas Egleston, defined engineering as "the science of making money for capital." A few years later, Henry R. Towne addressed the American Society of Mechanical Engineers (ASME) with a paper titled "The Engineer As Economist," in which he declared that the final word on an engineer's work "resolves itself into a question of dollars and cents." Towne continued to promote the idea, and in 1905 he told engineering students at Purdue that "the dollar is the final term in every engineering equation." Others presented the same message; in 1887, Coleman Sellers, president of ASME, said, "We must measure all things by the test, will it pay?"[8]

The role of engineer qua businessman led some mechanical engineers to an increased interest in the rational and scientific management of the workshop and factory. The engineer-manager viewed industry and business from his training as an engineer; he sought profits through factory efficiency instead of the traditional business pursuits of sales and was concerned with managing every element of production.

The development of industrial engineering was forecast in 1901: "The great opportunity for the engineer of the future is in the direction of management of our manufacturing industries . . . as competition grows sharper and greater, economies become necessary, the technically trained man will become a necessity in the leading positions in all our industrial works . . . He must be an Engineer of men and capital as well as of materials and forces of Nature." In the same year an article in the *Engineering Magazine* suggested that "a profession and curriculum of industrial engineering be organized similar to that of electrical and mechanical engineering."[9] But even before 1901 some schools offered classes that combined mechanical engineering and management science to teach what would become the principles and philosophy of industrial engineering.

Until the early twentieth century, the term *industrial engineer* was rarely used. In the late nineteenth century, it was an ambiguous term, usually referring to mill builders and production engineers, the inheritors of the early millwright's trade. The millwright, who had learned his trade through apprenticeship and experience, studied the power needs for individual factories and determined the size and type of power system to use and the kind of building to construct. That work would later become the province of the professional industrial engineer, who expanded on the millwright's job. Planning factories grew more complex as industry expanded and introduced new technology and as scientific management turned every facet of the factory into a planning problem.

The first men who called themselves industrial engineers were usually educated in traditional engineering programs. Many of them, such as F. W. Taylor, Frederick Halsey, and Henry Towne, had received training as mechanical engineers. Henry R. Towne, an early proponent of systematic management and a leader in industrial engineering, studied at the University of Pennsylvania and the Sorbonne and received a doctorate of commercial science from New York University. Charles Going, a lecturer at Columbia University and editor of *Engineering Magazine,* graduated from the Columbia School of Mines. Henry L. Gantt, a well-known protégé of F. W. Taylor, had been a graduate student of engineering of the Stevens Institute of Technology. Charles T. Main, owner of his own influential Boston consulting firm, received a degree in mechanical engineering from the Massachusetts Institute of Technology (MIT). Paul Atkins, a lecturer at the University of Chicago and a consulting engineer, graduated from Yale's School of Engineering.[10]

MIT was the first engineering school to offer classes that would train industrial engineers. In the mechanical engineering program in the 1860s, students in the fourth year could take a course that emphasized the coordination of major elements in manufacturing. A forerunner of later industrial engineering classes was one taught in 1866 which included "drawing of machines, working plans and projects of machinery, mills, etc." Though it might seem far from the total view of the factory found a few decades later, this class represents one of the earliest to include "mills" in the course work. The institute's catalogues show a gradual development in the mill or factory as an engineering subject. In the 1870s, MIT's mechanical engineering program began to emphasize topics such as "millwork," which "treats of placing machinery in manufactory, and the distribution, measurement, and regulation of force." By 1885, "during the fourth year the student is allowed to make a choice of one of the three following courses of lectures: First, a course on Marine Engineering; second, on Locomotive Construction; and third . . . on Mill Engineering." The design of the mill building proved to be one of the most popular topics in mill engineering. The 1892 courses included mill construction "studied together with the processes to be carried out in a cotton mill, so far as to enable the student to take up intelligently the layout of machinery to best advantage, including the planning of the power plant and the distribution of power, all leading up to the designing and building of the mill itself."[11]

The early course descriptions did not include management topics, but as the management movement developed, MIT students were exposed to the latest ideas. Beginning in 1886, the mechanical engineering program turned toward modern industrial engineering concerns with the introduction of special lectures by "gentlemen actively engaged in the profession," among them Charles T. Main, Edward Atkinson, Henry R. Towne, Joseph Stone, and F. W. Taylor. In 1894 the department added lectures on industrial management to the permanent curriculum and by 1898 offered courses on "Industrial Management and Foundations, the former involving a study of the organization and relations of the various departments of an industrial establishment."[12]

By the early twentieth century, other schools developed similar courses. Hugo Diemer taught shop management at the University of Kansas in 1901 and went on to begin the first degree program in industrial engineering at Pennsylvania State College in 1908. Diemer believed that economics, shop management, and the handling of labor were important to the education of engineers who would work in industry. He designed the program at Penn State to "fit men for business positions in industrial establishments." Other schools—Cornell, Purdue, Renssalear, the Carnegie Institute, Syracuse University, and the University of Pittsburgh—introduced similar courses and programs.[13]

Industrial engineering education paralleled the growth of management education in business schools. As the management movement took hold, business school faculty introduced the new ideas into their curriculum, and scientific management replaced an older model of business education. Some schools sought to develop experimental programs that would give equal emphasis to both business and engineering, giving new significance to the belief that engineering went hand in hand with business. In 1905 a correspondent to the *American Machinist* complained that too few young engineers exhibited the business knowledge needed to be good engineers, that they lacked "business knowledge and qualification." Such complaints were addressed by the new programs that would instruct students in business administration, commercial finance, and engineering. In 1910, the Carnegie Institute of Technology created its commercial engineering program to combine business and engineering. Harvard established a similar program "to study the principles underlying the modern organization of business and of recent applications of system," which would lead to "a study of the modern factory and of factory methods of production . . . with especial attention to questions of internal organization."[14]

In 1913, MIT began classes in its Department of Engineering and Business Administration which aimed to "furnish a broad foundation for ultimate administration positions in commerce and industry by combining with a general engineering training, instruction in business methods, business economics and business law."[15] Before setting up the program,

MIT conducted a survey of industrialists, businessmen, and engineers regarding the merits of the proposal. Thirty-five responded with enthusiasm, among them Charles T. Main, T. C. duPont, and E. H. Gary. In their letters of support for the new program, each described his frustration with young college-educated engineers. The president of Northern Pacific Railroad Company wrote: "It is a fact that we do not get enough men as engineers who display satisfactory qualities in a business way. They look at their problems purely from a scientific and engineering point of view, without enough consideration of the fact that somebody must provide the money for doing the work, and that work is not done simply to satisfy engineering pride and skill, but for the purpose of providing some facility that will be economical in maintenance and produce a real return upon the money invested."[16] Industrial engineer Charles T. Main replied that the proposed course would be "very desirable." He stressed the lack of knowledge of ordinary business methods on the part of most graduates: "most of the men, are unable to write an ordinary business letter, to say nothing of important reports or specifications, and if a little knowledge of this sort would be combined with the engineering courses, I think it would make the graduates better all round men." Another consulting engineer, Hollis French, said that "there is no doubt of the difficulty in obtaining the services of a competent administrative man. We find no difficulty in obtaining good engineers, but a good administrator, who is at the same time an engineer is a hard combination to find."[17]

By 1910 the profession began to take shape as the first formal university degrees were established. In 1912 industrial engineers received recognition from the ASME when the yearly meeting devoted an all-day session to industrial management. By that time industrial engineers had already gone a long way toward staking out their professional territory: according to Charles Going, industrial engineering consisted of "the efficient conduct of manufacturing, construction, transportation, or even commercial enterprises—of any undertaking in which human labor is directed to accomplish any kind of work."[18]

Despite the fact that the term *industrial engineer* linked the profession directly to industry, Going foresaw that the discipline would expand beyond industry, that their methods could be widely applied because they were scientific and would therefore enhance all work. One engineer wrote many years later, "Since industrial engineering is an analyzing, fact-finding, simplifying, measuring and controlling function, there is no phase of a business that cannot benefit from its use."[19]

The emergence of the industrial engineer—a hybrid of mechanical engineer, management engineer, and businessman—signified an expansion of engineering ideals of efficiency and precision. The industrial engineer did not design machines or products but instead planned factories and production systems and coordinated and oversaw all activities and operations in the factory. These new engineers left the "engineering of materials and enter[ed] the engineering of men."[20]

Industrial engineers writing about their history almost unanimously refer to Taylor as the "father of industrial engineering." Although Taylor played no direct role in building the professional industrial engineering programs, his scientific management studies played such an important role in the development of the field that some engineers even considered industrial engineering to be synonymous with scientific management. Even though Taylor began his machine shop reforms with experiments in high-speed steel, his followers defined their work in terms of management and economic efficiency rather than mechanical technology; as they built on Taylor's success, their vision of the "whole factory" went beyond Taylorism. Charles Going, a consulting engineer and the editor of *Engineering Magazine*, wrote *Principles of Industrial Engineering* in 1911, in which he described the discipline as the "formulated science of management." "Industrial engineering has drawn upon mechanical engineering, upon economics, sociology, psychology, philosophy, and accountancy to fuse from the older sciences a distinct body of science of its own."[21]

Despite the confidence of some, industrial engineers struggled with their identity during the early years of the profession. Where did they fit within traditional engineering practice, and how was their work different from that of the mechanical engineer? They sought to secure a place for themselves in industry, in part by becoming the most persistent publicists for the theories of efficiency and claiming management science as their own. Most importantly, they sought to legitimate their work by emphasizing its differences and distinguishing its contributions from those of other professionals. As they worked in industry, they used their new-found gospel to continue the pursuit of the nineteenth-century ideal of the rational factory. To those ends they discussed and wrote about their role in the industrial world throughout the early years of the emerging profession.

After its formation in 1917, the Society of Industrial Engineers provided a forum for those discussions. Within the organization, industrial engineers expressed differences regarding the profession. In his presidential address to the society in 1920, L. W. Wallace talked about the "confusion in the minds of many as to the real function of the Industrial Engineer." Discussions during the meeting confirmed Wallace's point; some argued that the industrial engineer was primarily a mill builder, confining himself to "industrial layouts, designs of mill buildings, and equipment, some appraisal work, and some engineering promotion." Another who supported the mill architect definition suggested that the role of the mill architect (plus production engineer) remained primary in 1920 and all other duties should fall within the category of counsel or advisor of industry. The majority, however, went beyond the mill architect definition and agreed with some version of Towne's 1905 description of the industrial engineer as the ideal manager who combined technical knowledge with administrative powers, "who can select the right man for the various positions to be filled, who can inspire them with ambition and the right spirit in their work; who can coordinate their work so as to produce the best final result; and who, throughout, can understand and direct the technical operation.[22]

Most of the new engineers thought of themselves as generalists. The industrial engineer considered problems of reorganization, personnel, equipment, buildings, and all the features of management and control in an industrial or commercial organization.[23] He was different from other engineers. He did not focus on the design of machines, the construction of public works, or any other traditional engineering operation. Each branch of engineering had its specific orientation, but the industrial engineer engineered work and capital instead of materials. Unlike other engineers, he based his work on a set of ideas about general processes rather than on a specific practice. This meant that his job more often involved planning than invention.

The one idea always at the center of the industrial engineer's work, the element that made all these activities one, was efficiency. The industrial engineer could never forget that his identity rested on the "science and art of developing the means of the most economical operation and control of a whole industry."[24] In addition, scientific management created an ideology of work—that any organization of people, whether for industrial production or business, education or health care, could be organized to be more orderly and efficient, thereby producing better results. The industrial engineer was thus the professional who reorganized work processes following the principles of scientific management. Most industrial engineers, however, worked in industry, and the confusion over their professional identity suggests that they did many jobs within the factory and that their responsibilities varied from job to job.

As they assumed much of the management of factories, the engineers' philosophy became increasingly influential, and as the rationalizing ideology came to dominate industry, each element of the manufacturing enterprise underwent reform. Factory layout and design became a major focus of industrial engineering. A modern engineer points to the factory as the starting point of the profession; the industrial engineer "became involved with the organization and design of the workplace and the flow of materials through the factory."[25] In their role as factory designers, industrial engineers extended the work of the early millwrights: they realized that through the design and layout of a building, they could move toward the ideal of the rational factory.

Industrial engineers and their employers left few records of exactly how and why they made specific changes in their factories, but we do have a record of their rationale as engineers saw it in their dozens of books and articles written between 1905 and 1930 describing the best way to build a new factory.[26] Guided by the principles of scientific management, they set out to engineer the factory, always working toward the goals of maximum profit and control. "Modern conditions," wrote one, "brought into organized industry a demand for systematic coordination of all factors which bear upon it."[27] The engineers wanted to organize the factory in the name of efficiency, so that the engineer-manager could control everything. In doing so, they considered every piece of machinery, every operation, and every movement in the plant.

Yet profits alone did not drive the engineers to recreate the factory; their faith in scientific management as an ideal also inspired them. Industrial engineers believed that scientific management, with its foundation in scientific rationality and uniformity, would improve society. The factory could serve as a model for other institutions—hospitals, business offices, and schools. Engineering was a process that could be applied as successfully to social problems as to technical ones.

While business economies continued to be a major motivation and the conception of the overall workings of the factory an essential feature of their work, industrial engineers focused on the factory building as a way to organize their ideas about efficient production. Through their work as factory planners, they reorganized every detail of the factory to increase productivity and reduce wasted time, energy, and materials; they introduced new management techniques, rationalized the labor process, and completely redesigned the factory.

The Factory

Redesigning the factory provided industrial engineers with one of their most important opportunities to rationalize American industry. As they planned new buildings and layouts, the image of the factory as machine guided them. The machine embodied rationality, a machine was predictable, controllable, nonidiosyncratic, easy to routinize and systemize. Like a machine, the rational factory could be set and adjusted; it could be turned on and expected to work at a predetermined pace in a predetermined manner. It would thereby correct the most serious problems in industry: production would be more predictable, efficiency would increase because there would be less wasted time and movement, the materials of production would be handled efficiently within the factory, and workers would have to work in a systematic way, giving up their old idiosyncratic habits that so bothered engineers and managers.

A prescient 1901 article in *Engineering Magazine* characterized industrial engineers' thinking about the factory and the worker. "Shop Arrangement As a Factor in Efficiency" laid out the problems of the 1901 factory and what the author, H. F. L. Orcutt, deemed the appropriate solutions—elimination of skilled workers and improved layout. The fitter remained a significant problem to the engineer in the 1901 machine shop. A skilled and well-paid worker, the fitter gave the final touch to all work to make sure it operated, or fitted properly. To the industrial engineer, he was a nuisance. Orcutt, like so many other engineers, advocated replacing fitters with machines to the point where "the skilful workers will then be small in number and their field will be confined to the making and maintaining of the plant instead of producing the article itself; the main body of workers in the modern shop will become little more than machine attendants." Once skilled workers were reduced to machine attendants, the industrial engineer could concentrate on the "modernly-designed shop," which should be designed to run like "a huge machine, of which the equipment and men are the moving parts."[28]

In 1903 the editor of the *Engineering Record* described the impact that newly graduated engineers had on plant design as they brought "skill and experience gained in bridge design . . . to the structural planning of the building." He emphasized his point by describing the old means of building factories. "It was formerly held by most manufacturers that they knew what they needed in the way of machinery, that the local millwright was competent to put in their power plant and a local builder to design and erect their mills . . . The local mason's buildings answered their purpose, although the floors sometimes sagged and the shafting was more often out of alignment, owing to the deflections of the structural framing."[29]

When he reflected years later on his early career, P. F. Walker wrote that as industrial engineers began to analyze production processes, "it became clear that existing factories were not arranged in a manner suited to [production operations]." With the knowledge that the new efficient production system needed a different factory, the designing and construction of industrial plants "assumed the standing of a separate phase of engineering work."[30]

By 1910 factory architecture had assumed new significance as fixed capital: as industries grew, large factory buildings became the major expense in setting up a manufacturing operation, adding to the importance of good planning and design. Setting up a manufacturing operation had clearly become too specialized for most owners and managers to do alone. With the recognition that the proper building could make or break a company, industrial engineers argued that the factory building's vital role in production required a trained professional to design it. After all, good factory planning required knowledge of the production process, business policies, and management, knowledge and experience that neither architects nor builders possessed. Engineers "especially trained in the planning and building of shops and factories,"[31] advised on site selection; size and design of the building; heating, ventilation, and lighting systems; shop floor layout; type of production organization and routing schemes for work flow; systems for materials handling; and just about anything else that had to do with managing production in the factory. Charles Day, author of *Industrial Plants*, described his work as "the latest manner of arranging and planning industrial plants."

He claimed that industrial engineering was "based on a logical scientific method of analysis which recognized not only all physical means available, but those more subtle factors having to do with the human element—men and women upon whom all industrial undertakings depend."[32]

In 1912, Henry Tyrrell, an engineer known for his early work with reinforced concrete, wrote *Engineering of Shops and Factories,* in which he described the changing nature of factories and explained the importance of the industrial engineer. "Those who were formerly content to carry on manufacturing in shops of the old type," he began,

> have long since discovered that the buildings themselves can be made one of the largest factors in economic production. The planning and arrangement of plants was formerly done by their owners or manager, who made little or no provision for their extension or development, and who considered that business success depended wholly on good management. It was then the belief that the buildings were of little importance, but it is now well known that they can and should be arranged and designed to facilitate production to the greatest extent.[33]

Writing about the early years of scientific factory planning, Harold Moore, in a 1925 *Mechanical Engineering* article, summed up the problems of poor factory planning as a simple and incontrovertible issue of insufficient space. "Many factories have modern equipment and efficient methods but an unbalanced distribution of floor space with some departments badly congested and others misplaced." He described lack of flexibility to rearrange or expand, constant problems caused by placing machinery in factories with columns too close together and ceilings too low to allow for overhead equipment.[34]

Industrial engineers considered the factory as part of industrial technology, not merely as a passive structure. Thus designing new factory buildings gave them control over production similar to the way designing a good machine controlled a specific operation. Industrial engineers began to call the building "the master machine," "the master tool," and "the big machine containing and coordinating all the little machines." They did not, however, believe that plant design was as perfect as machine design. Plant design needed further study, and engineers needed more experience to improve factory design practices: one argued that engineers' "vision must be broadened to embrace the idea of the plant as a working unit—a machine whose operation is a primary requisite to economical production."[35]

As they gained experience, industrial engineers realized that the arrangement of interior space in a factory influenced production efficiency in several ways. One of the most important was the impact on materials handling—the movement of all parts, raw materials, and in-process production through the factory. "The design of the building," wrote Charles Day, "should allow work to go forward as though the building did not exist at all." Charles Going suggested that the transportation of material was so important that it "may dictate the location and must control the design layout, and equipment of the plant." F. W. Taylor's protégé Henry L. Gantt, a strong advocate of the properly arranged factory, declared in 1912 that "two-thirds of all the difficult problems between obsolete inefficient management and the best possible one lies not in time study work, wage-payment systems, complicated functions of control, etc., but *in having all the material when you want it, where you want it, and in the condition you want it.*"[36]

In the late 1920s, L. P. Alford claimed that changes in "control of the flow of work" were the most important technical improvements in manufacturing during the preceding decade. The improvement, made through better layout and the use of conveyors and other devices, transformed production in that, by the time he wrote, the speed of production was almost completely determined by the speed of the handling system rather than by the speed at which workers wanted to work. These changes resulted in a reduction of supervision, extended the process of specialization, and made the "time of production definitely predictable."[37]

The new emphasis on the handling of materials allowed engineers to rationalize manufacturing further. Materials handling can, in fact, be considered a step secondary in importance only to the introduction of the manufacture of interchangeable parts and specialized machine tools. By the early twentieth century automatic machine tools were well developed, but production speed had reached a plateau because, as most industrial engineers agreed, the inefficient handling of materials constituted the greatest waste in manufacturing.

In poorly planned factories the right materials were not *where* they should be *when* they should be, a situation resulting in what the engineers called the hidden cost of materials handling. Excessive movement meant that more workers were employed to do the moving, and less was produced because an inefficient delivery system meant insufficient supplies at work stations. Some engineers argued that in poorly designed plants, handling accounted for greater labor costs than any other operation. Inefficient handling, they argued, was one of the most wasteful elements of production, and proper arrangement of departments could eliminate the unnecessary movement. By the second decade of the twentieth century, they proposed a variety of technical solutions such as craneways, gravity slides, elevators, and assembly lines, all of which will be more fully described in later chapters.

This approach to careful factory planning would address the labor problem as well. The well-designed building should "facilitate the economic management of labor." Labor costs constantly concerned the industrial engineer; they meant that "many a manufacturer is today facing the necessity of abandoning a plant which produces excellent goods, simply because of excessive labor costs." With proper engineering, workers would become part of the system, part of the great machine, and hence controlled like other variables in the factory. One efficiency engineer wrote, "I shall ignore the human element entirely as it actually exists in the shop and describe the people handling the operations as people who, whatever they may be outside the factory, are while in the factory simply animate machines . . . [trained] to do their work with all the precision of the most marvelous engine."[38]

To assure the highest possible productivity on the part of workers, some engineers stressed the importance of good visibility in the shop so that all workers remained in view at all times. Henry Noyes advised that one man supervise no more than three hundred feet of the shop floor. Other authors exhorted owners to build factories with open spaces and no hidden corners in which workers could hide as they shirked their duties. To that end factory designers should avoid the traditional building shapes that took the form of L's, E's, and H's. Concern with visibility and control of workers' movement around the factory also led to arguments against unnecessary walking or elevator riding. According to Hugo Diemer, "avoidance of unproductive travel demands a minimum of passageways," and those which were necessary "should be under the close supervision of watchmen who must note all wandering clerks and workmen, and who must be so informed as to the employees and their duties, that they may be able to observe and report illegitimate or aimless wandering."[39]

Proper arrangement of departments and machines was also important for the peaceful coexistance of workers, suggested some. "In certain localities it is not feasible to have union and non-union men working side-by-side in the same shop unless their work falls under widely separate heads . . . at these times union men refuse to work when non-union men are engaged in their midst." Charles Day gave the most specific advice when he cautioned manufacturers to segregate the foundry workers, often the most militant group, from employees of other departments "in order to gain adequate control of the labor situation." At the Wagner Electric Company in St. Louis, Missouri, foundry workers entered through a separate door and dressed in a separate dressing room from all other workers.[40]

The merits of specific design decisions such as factory size, shape, and number of floors became topics of debate among the engineers. Some argued for single-story plants and others for multiple stories: "Where the business is likely to grow," wrote one, "the disadvantage of a single-story building is that the organization becomes spread over so great an area that it cannot be properly supervised," whereas the multistory plant "lends itself more readily to expansion of business and unquestionably simplifies the general supervision of work." Others were convinced that workers were best supervised in single-story buildings with as few columns as possible blocking the view of supervisors. Some insisted that materials handling would be easier and faster on one floor; others said it was more efficient in a multifloor factory, that "in general it is easier to walk up a flight of stairs than to walk several hundred feet through departments." Finally, proper ventilation was difficult in the single-story building, and employees objected to "the closed-in effect."[41]

The multiple-story advocates apparently won in the early decades of the century, for few large single-story factories were built before World War II. Typically, one-story plants were built for the heaviest work—foundries, railroad car works, steel mills, machine shops, and forges. Producers of machine tools, firearms, clothing, shoes, and automobiles commonly used multistory plants.[42]

Ultimately even the best factory layout and design depended on sound construction. Not only was good construction a smart investment, but certain types of construction eliminated problems in the factory. For example, the introduction of reinforced concrete in the first decade of the twentieth century revolutionized the factory building. Concrete buildings required fewer columns; they generally had three bays instead of four, a factor that helped managerial visibility. Concrete almost eliminated vibration, which in multistory factories cost money through wasted energy. Vibration also caused significant discomfort for workers and damaged machines over time as well as causing machines and shafting to need constant realignment. Reinforced concrete buildings were stronger than wooden ones, allowing heavy machinery to be safely installed on upper floors. Concrete was also fireproof, a characteristic that significantly reduced insurance premiums, and many manufacturers trusted it so readily that they canceled their fire insurance. The "daylight" factory was also a function of concrete construction. A concrete factory did not depend on exterior walls for support and so allowed for small exterior columns, hence large areas were left free for windows.

Industrial engineering as a profession grew along with demands for greater industrial efficiency in the early years of the twentieth century. The new professionals employed an old metaphor—the factory as machine. They used it so often that we must read it as more than a colorful turn of phrase, especially when used by men who wrote straightforward technical descriptions of their work. The machine metaphor guided industrial engineers as they created a new kind of factory, one that moved closer to the eighteenth-century ideal of a factory that could run like a machine. The new factory could be organized in such a way that it provided machinelike predictability, forcing work to be done according to a master plan and thereby solving many problems of production. The lessons the engineers learned about planning the factory as a whole were so compelling that they were used in a wide range of industries and even outside industry.

The following example demonstrates the recognition of the work of industrial engineers as well as the usefulness of the ideas for institutions besides the factory. In 1913, when John R. Freeman, president of the Manufacturers Mutual Fire Insurance Company, was asked to help plan a new campus for his alma mater, MIT, he drew heavily on industrial engineering.

The institute had limited funds for the new campus, so Freeman argued that the best model to follow would be that of the industrial engineer, who combined economy and efficiency. Throughout Freeman's letters regarding the campus plans, he referred to factory planning practice. He proposed a campus design that he called "the factory plan," in which departments would be located in connected buildings rather than around the green spaces of the standard campus plan. "To anyone familiar with industrial problems it needs no argument or lengthy explanation as to the superior economy of the compact arrangement. Moreover, in our school of Applied Science we may very properly emphasize these economic features."[43]

Freeman argued that such a plan would result in cheaper construction (as proven in experience with factories) as well as more efficient use of the campus. He believed that most college campuses did not function well as teaching institutions because they were designed by architects to be monuments to rather than laboratories of learning. "I studied the problem," he wrote in his report to the campus committee in 1913, "from the point of view of an industrial engineer, who plans his buildings *from the inside,* and first of all arranges things with a view to moving the raw materials along the lines of least resistance."[44]

Although Freeman's design for the campus was altered, his use of the industrial engineers' rationale illustrates the legitimacy of the new profession as well as the acceptance of its approach in the larger engineering community. He demonstrated to a body of elite engineers (MIT faculty and alumni) that the principles of industrial engineering—of looking at the whole, considering the building to be the master machine—could and should be applied outside as well as inside industry.

The Human Machine:
Engineers and Factory Welfare Work

*It seems to me that the social side of the machinery needs lu-
brication as well as the physical side. The very complicated
mechanism of . . . labor must be frequently and carefully
oiled . . . Doubtless it costs something to keep the machinery
properly lubricated; it costs time and thought and patience,
and some money; but would it not pay?*
— WASHINGTON GLADDEN, *TOOLS AND THE MAN*, 1894

AFTER THEY HAD ENGINEERED THE FACTORY, INDUSTRIAL EN-
gineers turned their attention to workers, or the "human machine."
References like Gladden's, to the "social side of the machinery," were
common by the turn of the century, when engineers regularly expressed
their concern for the "cogs of the great machine" or the machine's "moving parts."[1] This conception of the worker fitted nicely into the industrial
engineers' rational factory framework: if workers could be thought of as
machines, then they could be studied, rationally selected, and even engineered to be better workers.[2]

In the second and third decades of the twentieth century, discussions
of the human machine grew in popularity among industrialists, industrial engineers, the increasing number of social scientists, and medical
professionals as a result of concern about industrial efficiency during
World War I. The Council of National Defense even issued advice, based
on Public Health Service research, on reducing industrial fatigue. Some
physicians, first concerned with fatigue, went further and developed
what they called "industrial medicine." Between 1910 and 1920 many
factories employed these doctors to work with personnel directors and
engineers to help make labor more efficient.[3]

In his 1918 book, *The Human Machine and Industrial Efficiency*, Frederic S. Lee, a physiologist at Columbia University, outlined "a science of the human machine in industry" developed so that "the highest degree of efficiency can be secured." His work, like many of the factory reforms described in this chapter, developed in part from "the unprecedented demand on industry" made during World War I. Lee described the usual complaints of the factory manager—high turnover, accidents, low output, soldiering—and his own prescriptions to treat them. While some of Lee's advice sounded much like that of any industrial engineer—hire qualified workers, pay a decent wage, employ scientific management— much of the book reflected his special training as a physiologist. He encouraged factory managers to study individual capacity for work and sources of fatigue, ensure that the work load was reasonable and would not result in excess fatigue, provide rest periods and fuel (food), and use industrial medicine.[4]

By 1927, when a consulting engineer, Richard Dana, published a similar book, *The Human Machine in Industry*, the idea of caring for the human machine had become a standard part of managing a factory. With the help of considerable research conducted over the nine years since Lee's book, Dana had developed some of Lee's ideas and included chapters on "fuel" requirements, cooling requirements, fatigue, and rest periods. The management of the human machine had become a science and a critical element of good factory management.[5]

For some engineers discussions of the human machine reflected their callous disregard for the humanness of workers. But for many engineers and other professionals concerned with industrial efficiency, such discussions prompted them to *remember* the human element at the same time as they sought greater control over work in the factory. Engineers addressed their concern for the human machine in two ways: by scientific study of the working conditions and by an appeal to workers' social and intellectual concerns through industrial welfare.

Scientific study of work environments led to changes in the factory such as improved lighting, heating, cooling, and the arrangement of work-stations. These studies examined the human machine quite literally—under what conditions can the human body work best (faster, longer, and more accurately). Industrial welfare explored the more human side of the human machine: what made "it" want to work harder. Although many different professionals addressed industrial welfare work, industrial engineers' involvement is interesting not because their contribution was especially significant but because they employed the new ideas as a balance to their efforts to rationalize the factory. They directed their work at old problems that took on new dimensions in the rational factory—efficiency and workers' productivity. In the rational factory these problems became social and psychological as well as mechanical; they were about workers as much as about machines.

Science of Work Environments

Most employers changed conditions inside the factory by applying "scientific" thinking to work. Engineers studied emerging research on industrial fatigue and experimented with the design of work stations and improved lighting, heating, and ventilation because they believed such improvements would result in better work from the human machine. In studying work environments, engineers generally went beyond simple concern for workers' comfort; they wanted to identify the perfect conditions for human productivity. Hugo Diemer suggested that even contented workers could not perform well in poor work environments: "poor air and insufficient light and warmth inevitably result in poor work as regards quantity and quality, even though the workers may be picked for their cheerful and sunny dispositions."[6]

Good light had always been a concern of factory owners, but until the second decade of the twentieth century, discussion of lighting focused primarily on windows. Getting daylight, considered the best light for work, into factories required innovations in building and window design. As late as 1915, Harry Franklin Porter, an engineer, wrote, "Window glass and white paint are better than electric lamps in the daytime for supplying factory illumination." Accordingly, industrial architects such as Albert Kahn enlarged factory windows by experimenting with reinforced concrete construction and changing the fundamental design of factory buildings, allowing the exterior wall to be almost all glass. The large windows installed in factories in that era concerned some employers, however. Some wanted translucent glass so that workers could not waste time by looking outside; others argued that a distant view gave the operative's eyes and mind an important rest from factory work.[7]

Artificial lighting became a factor in production as factory owners electrified their facilities and manufacturers of lighting systems produced lamps of better quality. Lighting engineers, as those who worked with lighting systems were called, decried the general inattention to lighting in American industry when, by 1909, a number of European countries had already established standards for their factories.[8] Lighting engineers urged better lighting for both its productive effects and for its impact on workers' attitudes.

Talks, articles, and book chapters on the benefits of good artificial lighting appeared in large numbers between about 1909 and 1929; they were so numerous that in 1914 one writer in *Iron Age* said that "in modern factory construction no object is probably given more attention than artificial lighting." In 1913 the Industrial Commission of Wisconsin was of the opinion that there was "a general awakening over the country among manufacturers to the economic value of conserving the human equipment in their plants, and there never was a time when so much attention was being given by progressive manufacturers to the subject of shop lighting." In a talk to ASME, L. P. Alford and H. C. Farrell explained the reasons why factory owners should invest in improved lighting: "greater accuracy in workmanship, increase in production, reduced accidents, lessening of eye strain, more cheerful surroundings and contentment of workmen, order and neatness in the care of the shop." Another engineer, stated the obvious: that "if the illumination is such as to strain [a man's] eyes, producing mental fatigue and nervousness, he will unconsciously be unwilling to work."[9]

Engineers working for a wide range of companies urged the introduction of good lighting because, as George Stickney of General Electric claimed, "workmen can do more and better work with an illumination of suitable intensity than with a weaker light . . . The illumination can increase his efficiency by making him at ease with his surroundings, or it can render him dissatisfied." According to Joseph Newman of International Harvester, "There is a tendency for the operatives to slacken their work as darkness approaches" because of the "inadequacy of artificial lighting . . . Lack of light has caused a large amount of waste of human energy and of material."[10] One might expect that by the time "darkness approached," workers were tired from a full day's work and for *that* reason were beginning to "slacken their work." That kind of problem, however, was much harder to engineer away.

The problem was that most factories were "designed for use by day." Factory design and organization usually did not accommodate electrical lights very well: "girders, belts, rafters, cranes, irregular ceilings, are the most common obstacles."[11] To make lighting matters worse, the electric lights installed in most factories before 1910 were of low wattage and hung in such a way as to illuminate only individual machines. They cast long shadows over the rest of the shop floor, making visibility bad and creating a generally gloomy work atmosphere that, some engineers argued, lowered productivity. New lights, equipped with reflectors to cast the light to a wider area, shone brighter and provided general illumination for an entire room rather than a single work station.

Like other factory improvements, good lighting would pay for itself. "Night production costs 20 per cent more than production in the daytime in an eastern silk mill . . . largely due to inferior illumination," reported *Iron Age* in 1911. A few years later, G. L. Chapin wrote, "Good lighting is, in reality, the acme of economy; it increases production, reduces spoilage of material and the percentage of 'seconds' and minimizes the number of preventable accidents." Engineers also suggested that better lighting made shop supervision easier by improving workers' attitudes and efficiency and visibility in factories crowded with machinery and belting.[12]

During World War I and in the years following it, engineers intensified their push for better factory lighting. The wartime labor shortage affected almost every factory in the country, increasing the need for efficiency and optimum productivity. Engineers argued that good lighting was more crucial than ever to conserve "the brain and physical ability of our men and women. Every factory under present conditions must secure maximum output with minimum waste of every kind." An *Electrical Review* editorial in 1918 observed that "the national crisis has made of improved factory lighting a most timely subject."[13] Even politicians were convinced of the importance of good factory lighting, and by the end of the war, several states legislated minimum standards for artificial lighting in industry.

An obvious relationship existed between good lighting and speed and accuracy of work, especially detailed work, but just how much difference did scientific lighting make to workers' productivity? This question had become so important by the early 1920s that in 1924 professors at the Harvard Business School tried to answer it in what became one of the most famous industrial experiments ever conducted in the United States. The Hawthorne experiments began with the seemingly simple act of

"WASTED"

Before—
After treatment with "Barreled Sunlight"

Wasted light and flaking paint! Do you realize how much *wasted money* it means?

Three thousand of the biggest plants in the country realize it, and they now treat their ceilings and walls with the finish that *increases daylight from 19% to 36%* and is permanent.

By using this finish, they help their workmen do more and better work; they decrease accidents; they *save as much as three-quarters of an hour electric lighting every day.* They save scaling and re-coating of cold-water mixtures, and flaking of paint into the machinery.

These plants have ceilings and walls that can be washed like a dinner-plate, and are thus kept wonderfully clean and sanitary.

The finish they use is "Barreled Sunlight"— Rice's Gloss Mill White—an oil paint made by a special process discovered and owned exclusively by the makers.

Repeated tests have shown, without a single exception, that Rice's remains white longer than any other gloss paint.

By the Rice Method, it can be applied over old cold-water paint. It does not flake or scale with the jar of machinery; it does not yellow like ordinary oil paints, and saves big money on painting because it does not need renewing *for years.*

"Barreled Sunlight" is also made as a Flat Wall Paint for office and hotel use. Sold by the barrel and by the gallon.

On Concrete Surfaces. Rice's Granolith makes the best possible primer for "Barreled Sunlight"—retarding the progress of moisture in the wall-Rice's GRANOLITH.

Write for our interesting booklet on factory lighting, "More Light," and Sample Board.

U. S. Gutta Percha Paint Company, 15 Dudley Street, Providence, R. I.

Reg. U. S. Pat. Office

THE ORIGINAL—THERE IS NO SUBSTITUTE

A promotional piece for Barreled Sunlight paint in 1917 played on the fear of waste and lively interest in good industrial lighting. *Factory*, Dec. 1917; photo courtesy of Hagley Museum and Library, Wilmington, Del.

measuring the effects of lighting levels on productivity. The experimenters, imbued with the modern belief in scientific techniques and principles as the way to resolve all problems, designed a careful study that would yield detailed knowledge of the relationship between work and lighting. The Hawthorne experiments became famous, but not for what they discovered about lighting. Researchers found that productivity within the small experimental group improved with better lighting systems *or* worse ones; they concluded that the special attention of the scientists and the comaraderie of a small group had a greater effect on performance than the physical working conditions. The experiments explained more about problems in social science research methods and worker psychology than they did about factory environments, and they launched many subsequent investigations.[14]

Industrial engineers and factory reformers also believed that, like lighting, air quality in the factory affected workers' ability to perform, and good heating and ventilation became a factor in the proper functioning of the human machine. Though industrial engineers gave it much less attention than lighting, proper air quality became a standard prescription for good factory planning, and some even considered it the most important element of the work environment. "The very first essential to the solution of the humane elements in the machine-shop problem is clean and well-ventilated buildings," wrote John H. Patterson in 1901. Concern for good air quality went beyond a humane concern for workers; engineers thought that, like so many other considerations, it paid off in higher productivity. "The money invested in those items relating to the bodily comfort of the operatives will be found to pay a substantial dividend in the reduction of absences from sickness and in greater physical vigor and mental alertness."[15]

Specific health problems were linked to poor environmental conditions in the factory. One engineer was of the opinion that when the factory operative worked in extreme heat throughout the winter months, "he works listlessly, he half accomplishes his task, he breaks and wastes property and the material entrusted to his care." Furthermore, when exposed to inappropriate temperatures "his vitality [is] lowered . . . and he falls a prey to minor illnesses . . . absences . . . and tuberculosis." He concluded that "conditioning the air so that the human machine may work under the most favorable conditions is one of the chief elements of industrial efficiency." The most favorable conditions were, according to another engineer, "the weather conditions on a mild spring day," and the industrial plant should try to "maintain a temperature of 65 degrees . . . with relative humidity of 40%–50%."[16] One book chapter titled "Cures for Factory Absences" pointed to harmful factory conditions. For example, heat prostration might be alleviated by hosing down a tin roof or by removing false ceilings, and lung troubles could be reduced by constructing a roof over a cold alleyway between heated rooms.[17]

Controlling air quality was a far more difficult task than improving lighting. The requirements for consistent temperature and humidity throughout the year were more complicated than providing good light. Heating cold air was the easier task. Some factories used the exhaust heat from steam engines to provide warmth to their workers in winter months; others used live, or direct, steam heat. Cooling hot air was harder. Some factories installed tanks of water, which along with fans helped cool the air. Fans were recommended for use in most factories. They not only distributed heat but also helped to cool in the summer, and if properly installed, could draw in fresh air. Some engineers suggested that the human machine could function better in an environment with constant air movement. As early as 1902 air conditioning began to appear in factories. An expensive system to install, early air conditioning addressed humidity factors rather than temperature, though reduced humidity certainly improved human comfort.[18]

Quite unlike other considerations for the human machine, the condition of the air, especially humidity, also affected certain manufacturing processes. The textile industry required a certain amount of humidity in the air to ensure that threads did not break during spinning and weaving operations. The candy industry, especially companies making chocolate, found that their product was so adversely affected by summer humidity that they often closed down during the hottest months. Others, such as paper mills, dye houses, foundries, tobacco processors, pasta makers, and flour mills, were similarly affected.[19] Thus by addressing air quality, engineers often alleviated both production problems and those of workers.

Industrial engineers also developed a science around the study of work motion. Frank and Lillian Gilbreth, the most famous of the motion study pioneers, studied the way workers moved through the course of performing their jobs, so that work could be designed to reduce fatigue and increase efficiency. As with other factory reforms, the goals of motion study were increased productivity through more efficient work techniques and reduced fatigue. Prescriptions for workers' movement which resulted from the motion study analyses evoke the image of the machine: each human machine working exactly like the one at the next work station.

An early follower of F. W. Taylor, Frank Gilbreth introduced his version of time and motion study in 1912 in his *Primer of Scientific Management.* Throughout his career he focused on motion study over other elements of scientific management and, with his wife Lillian, developed sophisticated techniques to study work. The pair pioneered the use of photography to study workers' movements—by attaching lights to a worker's hands and then taking motion pictures in a darkened room, they created a record of work motion. The lights allowed the researcher to note the exact series of motions, something that was impossible by simply watching.[20]

The Gilbreths went beyond the technical details of motion study; they also introduced early principles of industrial psychology and advocated special training in schools to prepare young men and women for factory work. They proposed that students be made "finger-wise," meaning that the muscles of the hand be trained so that "they respond easily and quickly to demands for skilled work."[21] The Gilbreths and their fellow engineers wanted to change the meaning of skilled work; they wanted the term to refer to the operation of a factory machine rather than to the ability to create something or exercise judgment about materials and processes which characterized the artisan.

When industrial engineers asked how they could "oil" the human machine, how they could get more and better work out of human beings, they did not seek simply to *make* workers work harder. Nineteenth-century paternalists and twentieth-century "welfare secretaries" worried about workers' attitudes and sought to make them *want* to work harder by treating them like "family." The industrial engineer, however, used the science of work environments to create optimum working conditions. Here were factors affecting production of which the worker was unconscious. To improve productivity independently of workers' attitudes was preferable to depending on their good will. If one relied on providing good conditions, perhaps the personal could be overlooked. To engineers, the study of worker physiology approached more closely the image of the human machine that worked along with the master machine.

The Industrial Betterment Movement and Welfare Capitalism

Washington Gladden's expression of concern for the human machine which opened this chapter was part of the industrial relations movement that began at the end of the nineteenth century. The movement, also called "industrial betterment," "factory welfare work," or "welfare capitalism," began in the late 1880s and ended during the Great Depression. It attracted a surprising mix of supporters, from liberal reformers to industrial engineers and factory owners. Though most scholarly attention has been paid to the welfare capitalism of the 1920s, it was the early part of the movement that addressed the human machine.[22]

Welfare work, as it began in the late nineteenth century, resembled the paternalism of the earlier part of the century in which the company regarded workers as symbolic children who needed to be watched over. Companies offered, or sometimes required employees to use, company housing, schooling, churches, and other extrafactory services. Much of the paternalistic attitude remained in later welfare programs.

Factory welfare work, as promoted and practiced by the engineers, centered on the seemingly conflicting ideas that the worker was a cog in the great machine but also had to be treated as a human being. Welfare secretaries also used the "oiling the machine" metaphor—welfare work was to the worker what oil was to the machine. If he or she could be made happier, more fit (physically, intellectually, and morally), and more secure, the company could expect greater productivity, less labor unrest, lower turnover, and better all-around labor-management relations. In other words, through industrial welfare the factory owner and manager hoped to alleviate much of the labor problem.

Indeed, the labor problem, which had been ever present to nineteenth-century industrialists and engineers, had not gone away; but it was redefined. As the focus of so much of the scientific management writing, it became a primary concern for industrial engineers and a regular topic of discussion in engineering journals. By the beginning of the twentieth century, engineers defined the labor problem in several ways: getting workers to work the one best way, as per Frank Gilbreth; the need to reduce skill requirements; the threat of trade unionism; the high wage rate; and high turnover.[23]

Welfare work could help to solve some of the labor problem. H. F. L. Orcutt wrote in 1901, "The most-productive and skilful workers will seek shops which are clean, warm, ventilated, and well lighted, with the best sanitary arrangements and facilities." He went so far as to say that most of the "so-called labor troubles are due . . . to the fact that so many of the masters build shops and run them absolutely ignoring the comfort, interest, and welfare of the most-important factor of their business—their workmen."[24]

But the concerns of some industrial engineers, such as Charles Going, went beyond the labor problem and factory operations. Going was concerned with a matter he felt so important that he gave it regular and lengthy treatment in engineering journals and books—the reorientation of the United States into a placid industrial society. Going recognized that "mechanical production is working a revolution in economic affairs . . . it is also working a revolution in social conditions and relations." He saw a problem growing along with the growth in industry: "the close personal contact between the proprietor and workman, which belongs to the domestic and semi-domestic era of manufacturing, must give way to . . . a new order."[25] He and his fellow industrial engineers sought the new order, in part, through industrial welfare.

Industrial welfare work, as an integrated system, began with "model company towns" like George Pullman's, named after himself and his company, and N. O. Nelson's LeClaire, both in Illinois. These towns were built by industrialists who believed that the proper social and physical environment would encourage the kind of personal and moral development that made for loyal, hard-working employees. Pullman's experiment, begun in 1881 and steeped in a rigid, old-fashioned paternalism that dictated all behavior in the town (workers could not own their homes, no alcohol was allowed, and the company planned everything down to gardens) resulted in dramatic failure. In 1894 his workers staged a strike of such magnitude that it shook the nation.[26]

Nelson started his profit-sharing system in LeClaire in 1886 and met with greater success. Whereas Pullman sought control over every element of his community, Nelson described his community as cooperative. Nelson's attitudes reflect the more modern ones of the industrial betterment movement. By 1915 his company and town had already demonstrated his successes. "The problem of increasing importance to the factory manager is his relation to the employees." The labor problem, as

Roof Garden for the Women and Girls of the H. J. Heinz Company, Pittsburgh. Many early-twentieth-century companies provided dining rooms and "rest" rooms as well as libraries and reading areas for female employees. From William Tolman, *Social Engineering* (New York: McGraw Publishing, 1909); photo courtesy of Hagley Museum and Library, Wilmington, Del.

Nelson defined it, consisted of the need "to reduce turnover, insure steady output and prevent interruption of production through differences between yourself and the men . . . the existing labor population brought up under the old methods of production do not take kindly to the factory system with the curtailment of their freedom."[27] They had to be persuaded to like the new methods.

Dozens of companies introduced welfare programs. In 1905 the Ludlow Manufacturing Association in Massachusetts built an employee clubhouse that included a theater, gymnasium and dance hall, poolroom, bowling alley, card and smoking rooms, baths and swimming pool, and locker rooms. In 1911, John H. Patterson, owner of the National Cash Register Company, built baseball diamonds, tennis courts, a dance area, and a golf course for his employees in Dayton, Ohio. Between 1912 and 1925 the U.S. Steel Corporation spent over $158 million on its welfare program to provide playgrounds, schools, clubs, gardens, safety features, accident relief payments, and pensions.[28]

Believing that good housing would engender stability, some companies developed simple programs of supplying housing, either for rent or for purchase at cost. After World War I some employers expressed their belief in the power of industrial welfare, especially home ownership, to stop the spread of Bolshevism.[29] Company-supplied or built housing was a mixed blessing, however, for the employee. Though the housing was inexpensive, a change of jobs meant eviction from rented accommodations or pressure to move out of a house under mortgage to the company. Housing thus constituted a significant inducement to remain in the employ of the company that owned it, reducing turnover for the employer.

Other factory owners went farther. Though few built entire towns as Nelson and Pullman had, many developed elaborate programs with the help of welfare secretaries. They offered amenities unrelated to the direct work of production: gardens, dining rooms, gymnasiums, swimming pools, theaters, dance halls, libraries, clinics, commissaries, and locker rooms. Outside the factory, in addition to housing, a company might set

Setting Up Drill at the Thomas G. Plant Company. An extreme example of the healthful activities companies organized in the factory welfare era. From William Tolman, *Social Engineering* (New York: McGraw Publishing, 1909); photo courtesy of Hagley Museum and Library, Wilmington, Del.

up a school for children of employees and facilities for outside recreation. The most frequent addition to the factory itself was the workers' dining room. Formerly, workers carried their lunches or bought them from vendors who wheeled sandwich carts around the shop floor and ate at their machines or outside on the factory grounds. Bars and cafes sprang up near many factories, providing another, more attractive, option; many bars offered free food along with purchased drinks.[30]

Owners built dining rooms for several reasons. Factory inspectors pointed out the health hazards inherent in the practice of eating on the shop floor near the dirt, chemicals, and other dangerous materials of production. The general absence of washing facilities in many plants exacerbated the problem. Managers did not like workers to leave the factory for lunch, especially when the men were likely to patronize the neighboring saloon. Not only did employees return late from outside lunch trips but their lunchtime imbibing hurt their productivity. The saloons created another, less obvious, problem for management—they often served as gathering places of union organizers.

Welfare secretaries convinced employers that their employees could work better with a healthy, hot lunch provided in a peaceful, wholesome atmosphere where they could experience a real break from work. The effort and expense in providing dining rooms varied widely. At the Iron Clad factory in Brooklyn, the "dining room [was] the most important feature." A large room with many windows, each holding a flower box, and a ceiling covered with grapevines, the dining hall was arranged like a restaurant, with small tables. Waiters in white uniforms served workers. On a more modest level the Waltham Watch Company of Waltham, Massachusetts, gave employees the use of a plain room with small tables. The company supplied a counter "for the sale of simple forms of food" and facilities for heating food and coffee. Most companies that built dining rooms also supplied some form of subsidized lunch. The Pierce-Arrow Company furnished a hot lunch for fifteen cents; National Cash Register employees also paid fifteen cents for a hot lunch in "Welfare Hall"; the Plymouth Cordage Company sold a "substantial dinner" for ten to twelve cents; and the Natural Food Company of Niagara Falls gave their employees a free lunch.[31]

In addition to dining rooms, sometimes owner-built sitting rooms, smoking rooms, and libraries provided workers with quiet retreats to "refresh and educate themselves" during breaks and after working hours. Sitting rooms supposedly offered women, and sometimes men, a haven from the busy factory. Some companies such as the Shredded Wheat Company, National Cash Register, and National Biscuit "required" their workers to take a ten-to-fifteen-minute break each afternoon. Many allowed no breaks, however, and one must presume that workers could use the rooms only at the end of the workday. These "quiet" rooms characterized welfare programs at companies that employed a predominantly female work force. Many managers argued that women would marry and leave factory work, which meant that they did not need long-term programs such as pensions and health insurance. Employers assumed that men would be more concerned about personal security—pensions, profit sharing, housing, and job security.[32] Welfare features for women, consequently, cost the company less and meant less to the women.

Recreational facilities were among the most interesting additions to factories under welfare programs. Guided by their belief that most workers were of weak character, welfare secretaries devised recreational programs to influence employees' behavior outside the workplace. Industrialists feared that leisure threatened work values, that workers would not use their increased leisure in wholesome, healthy pursuits. The programs aimed to counter the influence of unhealthy commercial leisure such as the dance halls, saloons, pool halls, and amusement parks that appeared in any industrial town or city. Many industrialists and other middle-class citizens believed such entertainment was bad for industrial morality. Company recreation would increase efficiency by keeping the workers healthy and sober, and, perhaps more importantly, it would build character and create a team spirit among them at the same time as it instilled company loyalty.[33]

Factory owners sometimes went to great expense to make company recreation attractive and convenient. Several companies built swimming pools inside their factories. Others built bowling alleys and installed pool tables in a club room. Some built elaborate clubhouses for employees' use. Still others added auditoriums and ballrooms to the factory. Less striking were ball fields and picnic areas on the factory grounds. Employers felt the expenditures were justified because good recreation, like other welfare work, supplied maintenance for one of the most important machines in the factory—the worker.[34]

Engineers and "Rational" Welfare Work

After the factory had been reorganized, new machines installed, and work arranged according to time study, industrial engineers began to realize that productivity still was not as high as they wanted it to be. And to add insult to injury, so to speak, they came under criticism from other professionals—the social engineers as well as politicians—for forgetting to pay attention to the human factor.[35] They had been concentrating on the technical details of production and overlooked the human element. In their planning, they had forgotten that production depended on human will as well as technology. Engineers could design the ideal technical production system, but if workers did not want it to be productive, the system would not be productive. As work became increasingly rationalized, workers' individual output was expected to grow. But as their jobs changed, allowing them less control over how they performed their task, workers refused to speed up their work to the extent the engineers believed possible. There was no way around it: without motivated workers, no amount of planning would attain the desired rates of production. This problem led to the combining of scientific management and welfare work.

The industrial welfare movement included a variety of responses to the transformation of work in the United States in the last decades of the nineteenth century and the early decades of the twentieth. Industrial engineers along with new professionals who called themselves social engineers worked together to put an end to any question about the motives behind their welfare work.

In a 1901 article entitled "The Social Engineer, a New Factor in Industrial Engineering," William Tolman introduced what he called social engineering.[36] It was not really engineering, though it shared some of the concerns of industrial engineers. The new profession reflected both the growing interest in social reform within industry and the strong influence of engineering as a profession and ideology. Because of its growing significance, engineering ideology influenced many professions, and the term *engineering* was often used for nontechnical jobs.

William Tolman, the "father of social engineering," and the Reverend Josiah Strong, a minister of the social gospel, established the American Institute of Social Service in 1902 for the purpose of "social and industrial betterment." The institute gathered information on employers' welfare work from around the world to disseminate to American industrialists. Based in New York City, it encouraged industry to use its services in the interests of workers, society, and the company. Some of the largest firms in the country appear on the lists of the institute's clients—Prudential Insurance, General Electric, McCormick Harvester, Sherwin-Williams, H. J. Heinz, and National Cash Register.[37]

Under the auspices of the institute, Tolman wrote his *Social Engineering*, in which he described his new profession and its role in society. The social engineer would help to bring about "industrial peace and contentment" *and* help make the firm more efficient. Industrial peace would come in part when the industrialist established a connection between himself, his staff, and "the rank and file of his industrial army." The social engineer would provide this connection because the "industrial army" had grown too large for the employer to do it himself. Employers were afraid that workers mistrusted the motives behind welfare work; the social engineer could mediate any such problems between employer and employee.[38] Workers had, indeed, begun to question factory welfare work; they argued that welfare features simply drew attention away from low wages and long hours. These critics suggested that if an employer paid fair wages for reasonable hours, he would have no trouble with workers' loyalty.

Obviously social engineers were not "real" engineers; they manipulated people rather than machines. They did, however, adopt the basic premise of the industrial engineer: organizing the factory rationally would make workers work as they should and thereby increase productivity and the general prosperity of the society. Social engineers talked about efficiency promotion (that is, getting workers to accept efficiency work), hygiene, safety, security, benefit associations, housing, pensions, employee thrift, education, recreation, and community betterment. Like the industrial engineer, the social engineer did not consider any of this work to be altruistic. She concerned herself with improving the factory's working environment only to the point where it would be profitable.[39] As with everything else in industry, environmental improvements were judged, in the end, by the degree to which they enhanced profits. There was "little room for sentiment; the ordinary employer demands a cash equivalent for each dollar paid out."[40]

In the early years of the twentieth century, industrialists realized that they needed a system to manage the work force just as they had a system to manage the technical side of production. Turnover, threats of strikes, and workers' noncooperation seemed to be the final hurdle to achieving a smooth and, more important, reliable flow of production. Thus welfare work became part of a broad-based strategy with which industrialists addressed the labor problem. Welfare capitalism, especially in the period 1910–19, was an effort to stabilize the working class; to stop (or at least curtail) the radicalization of workers.[41] These ideas became so important to industrial engineers that "industrial service work" was introduced into industrial engineering training programs; by 1916 over 150 programs included some kind of course to teach the engineers about the latest ideas on how to get workers to work along with "the machine."[42]

Predictability was an important part of the new industrial system—the ability to keep contracts and the faithful delivery of promised goods were critical to the expanding industrial system. Workers' cooperation affected predictability just as it did productivity. Washington Gladden realized this connection: with good relations the employer "can make his contracts with confidence; he will feel well assured that he will not be interrupted by threats of a strike when the tide of business turns toward prosperity."[43]

Industrial engineers' writings illustrate their concern with the human element in manufacturing. During the first thirty years of the century, books on factory planning and design appeared with chapters or sections extolling the effectiveness of "welfare features" in the factory.[44] These sections proposed a combination of managerial strategies—the concurrent use of scientific management and welfare work. Some writers explained in detail how and why welfare practices contributed to productivity. Others treated industrial betterment in a perfunctory fashion, implying that they grudgingly accepted the necessity of the new elements of factory planning. Generally, writings of industrial engineers reveal that their purpose in using welfare work was to produce efficient workers by treating "the operative as one of the factors of production whose efficiency should be raised to the highest pitch."[45]

By the 1920s specialization of professionals in the factory was as visible as the specialization of production operations. The lower-paid social engineers assumed principal responsibility for welfare work as industrial engineers began to relinquish their role in that part of factory planning. Harrington Emerson, a spokesman for industrial engineers since the earliest days of the profession, said in 1922:

When I was much younger I thought that the function of the industrial engineer was to spread himself all over the map. He was concerned with the health of the operators, with their education, their morals, and their happiness—all that was part of the function of the industrial engineer. As I went along I discovered it was none of his business whatsoever; that he was there to secure industrial competence . . . Happiness, of course, is a great thing for the human race, but it is not the business, as I see it, of the industrial engineer to take up the subject of happiness.[46]

Industrial engineers played an important role in the growth of industrial welfare programs; in many ways they legitimated it.[47] Not willing to give up any part of their control over what happened in their factories, industrial engineers developed their own brand of welfare work. Engineers took what some viewed as an idealistic, humanitarian idea and turned it into a practical tool to increase production; most engineers were, in fact, quick to deny any humanitarian or philanthropic motives. Hugo Diemer, a noted industrial engineer, wrote:

There are two reasons for employers giving workers good air, good light, proper temperatures, safe and comfortable working conditions, good lunch facilities, and encouraging athletic and recreation facilities. The first is that better health and better attitude . . . help to produce the largest possible output at the lowest possible cost . . . The second reason is based on the idea that industry and society are interdependent, and that industry and business have certain obligations to society in the way of helping to develop and maintain a healthy and intelligent citizenship . . . It will be noted that neither of these reasons involves anything in the way of philanthropy or paternalism.[48]

Despite the efforts of welfare workers to conceal it from them, unions and many individual workers understood that welfare work was motivated more by profit and control than by altruism and social justice. Understandably, unions criticized welfare programs; authors of welfare proposals made no secret of their intentions to keep unions out of the factory. Managers hoped that industrial peace and stability would follow the introduction of welfare programs, eliminating union appeal. For many employees welfare capitalism meant that a major issue in the labor movement—working conditions—ceased to be a problem. For others the superficial efforts of welfare work only hid the fundamental inequities of

the modern industrial system. Labor leaders argued that welfare practices lulled workers into inactivity and replaced the more important
hour and wage reforms.[49] Though some companies included wage bo-
nuses in their welfare programs, many others offered only nonmonetary
improvements, which gave employers more control over workers than
they would have had with higher wages. With better wages employees
did whatever they wanted with their money. With welfare programs,
employers determined how the money was spent. As one critic wrote in
1926, "the welfare of the workers is constantly becoming more and more
dependent upon the good-will, success, and prosperity of the particular
industry in which they are engaged."[50]

Welfare work had profound implications for the workers' community
as well as the workplace. If the factory programs succeeded in appeal-
ing to workers, the community would become dependent on factory-
supplied recreation and social services. Dependence would grow as other
sources of recreation and services disappeared through disuse. Some employees foresaw the potential dangers in allowing the factory to provide
for their social life as well as their employment, and they opposed each
new addition in the factory's welfare work. At the 1928 convention of the
Amalgated Clothing Workers of America, one worker proclaimed: "Welfare work is a deadening anesthetic. It is Delilah's method of robbing
Samson of his power . . . It puts the employer's collar on the worker . . .
It chains him to the factory not only as a producer of goods but also in
every other respect. Even his recreation is handed to him at the factory,
in the factory atmosphere, and with his employer's label. Under the
welfare system the worker is a 'factory hand' even while singing and
dancing."[51]

Some workers directly rejected welfare efforts. In one case, rather than
use a library the company built, employees collected their own library
fund. Workers often refused to take advantage of subsidized meals; when
asked why he did not buy the two-cent coffee, one replied that he was
"afraid that if he took two cents' worth of coffee he would be expected to
do seventeen cents' worth of work for it."[52]

Despite some criticism by workers and strong opposition from organized labor, welfare work enjoyed significant successes from 1910 to
about 1930. Welfare programs suffered during the depression of the 1930s
for two reasons. Massive unemployment brought industrialists a large,
willing pool of workers who had no intention of leaving a job of any kind

in search of another—the turnover problem was solved. Financial hardships of the depression necessitated cuts, and welfare programs were easy to eliminate.

To some, welfare work, or welfare capitalism, seemed little more than age-old paternalism used by an increasingly sophisticated industrial system. While it mimicked parts of paternalism such as concern for the worker's life outside the factory, it did much more than bind the employee to the company: it helped to make the employee into a better, more reliable worker. Industrial welfare work made the concern for the worker part of the industrial bureaucracy under which, regardless of the claims, workers were treated as part of the grand machine.

Like so many other reforms related to the progressive movement, welfare work came under the supervision of experts, newly trained in social engineering. Social engineers, welfare secretaries, and often the industrial engineers themselves formalized the relationship between the worker and the company, a necessity born of the increasing size of the work force. With thousands of workers in their factories, owners and managers could not practice the old style of paternalism under which they claimed to maintain personal contact with employees. Not only was it impossible for an owner to know employees personally, but he did not want to know them; to know them might also mean to feel a personal responsibility for them. And after all, personal feeling toward the "cogs" of the great machine would be unproductive.

Modernizing Factories in
the Early Twentieth Century

THE INNOVATIONS OF THE NINETEENTH CENTURY—SPECIAL-purpose machines, high-speed steel, scientific management, and others—made the running of a factory increasingly complex. Early in the twentieth century, industrial engineers, having learned so much from nineteenth-century industrial innovations, better understood how to make production run more efficiently. In the first two decades of the century, they made significant progress toward building a rational factory. The authors of articles in journals such as *Iron Age, Engineering Magazine, American Machinist,* and *Factory* laid out the major concerns that a good building could address: reducing the expense of materials handling, eliminating nonproductive labor, increasing productivity, decreasing downtime of "high-priced" skilled workers, and ensuring regularity and predictability of the production schedule. The most striking and consistent feature of discussions about new factory buildings was the concern over making the building fit production rather than organizing production to fit into a standard building. A clear articulation of that point was made in a 1908 *Iron Age* article that described the new plant of the Kennedy Valve Manufacturing Company. "The machines were grouped in plans, as required for economical production, and *the buildings drawn around them.*"[1]

Slightly more than a decade earlier, in 1896, industrial journalist Horace Arnold had written a series of four articles for *Engineering Magazine* which discussed shop location, shop construction, shop plans, and management. Arnold was ahead of his time in many of his observations. He wrote that "the form of machine-shop buildings has a very great influence on the labor cost of machine work. . . [the] well-equipped and well-managed shops can be handicapped by the mere shape of the buildings to the extent of ten per cent."[2] After touring many shops and talking with their owners and managers, he advocated large, rectangular, single-story shops with as few partitions as possible (partitions hampered the supervising of workers). Arnold went on to describe the features that characterized the best machine shops of 1896. They included electric traveling cranes to help move heavy materials around the shop and roofs that let in daylight, especially saw-tooth roofs. Many of the best shops included balconylike "gallery" floors built beside the high open machine shop. What Arnold saw in 1896 was the beginning of serious change in the way owners and engineers thought about the construction and organization of factories.

The changes in three technologies—materials handling systems, reinforced concrete, and electricity—would open the way for engineers to move closer to the rational factory.

Materials Handling

While industrial engineers argued that many problems could be resolved with proper factory design, the problem they most often addressed as they talked about new factories was the handling of materials. In fact, improvements in that area would, they suggested, not only address the elimination of nonproductive labor but also would help manage the remaining workers. Efficient handling would keep machine operators at their stations and keep them well supplied with parts and materials, preventing them from walking away from their machines in order to restock their stations.

Some nineteenth-century industries had recognized the important role that materials handling played in production, but the larger manufacturing community came to the same realization only during the first decade of the twentieth century. A 1905 article explained that the capital invested in manufacturing operations had grown so large that every small saving counted: "It was not a struggle for minutes, but for seconds"

A traveling crane lifts a locomotive under construction at the Baldwin Locomotive Works, Philadelphia, 1896. Before Ford, traveling cranes typically
lifted heavy objects rather than moving materials within a factory. *Engineering Magazine* 11 (1896): 277; photo courtesy of Hagley Museum and Library.

and mechanical transportation in the shop would result in significant
savings. The aim of new manufacturing processes and management
techniques was to decrease "prime labor costs" and increase "output on
plant investment." Anticipating Henry Gantt's more prominent state-
ment (cited in chapter 2), the author stressed that mechanical transportation was important "to get what you want, where you want it, when
you want it."[3]

Soon after 1910 the industrial community agreed on the vital role of
materials handling in the factory. Engineering journals published article
after article on the topic, and it is clear that factory engineers agreed that
"one of the greatest fields for improvement . . . is that of internal transportation." One author wrote that engineers had done a good job of
improving the "productive" departments in factories: "time study and
scientific arrangement of machinery have made a great increase in the
output of their plants." But "a study of the non-productive side of manufacturing has brought about improvements which also show a great
savings . . . One of the largest non-productive items in a plant is the
handling and moving of materials."[4] Others supported that position: one

author wrote that "while the possibilities of savings [by improving machine operations] have by no means been exhausted, engineers engaged in management work are now finding it profitable to turn their attention to other phases of factory operation [especially internal transportation]." Another claimed that "one of the largest items of expense in the operation of the manufacturing plant is the cost of moving material through the shop while it is in the process of manufacture."[5]

These engineers recommended changes more fundamental than simply replacing laborers pushing hand trucks with electric trucks and cranes, changes that reflected a significantly different way of thinking about factory production. Many of the engineers writing about handling innovations talked about not just improving handling but eliminating it. They described an early version of "just-in-time" manufacturing. "The governing idea in all studies for the cutting down of unnecessary movements between machine operations is that the final position of a piece in one operation should be the initial position for it in the next operation, and there must be no rehandling between operations."[6]

"The ultimate aim of nearly all revised shop methods," read a 1911 *Iron Age* article, "is to keep the machine tools operating as continuously as possible." Efficient handling ensured the greatest output per machine because the well-paid machine operator used all of his time running his machine instead of looking for parts.[7]

Industrial engineers were experimenting with a range of techniques to eliminate the "high cost of materials handling." The Hydraulic Pressed Steel Company in Cleveland, "realizing that it costs money to set material down and to pick it up," introduced a system of handling that allowed no material to "touch the floor from the time the raw material . . . is taken into the plant until the finished product leaves the shop. The steel sheet or partly finished part will be taken from one truck at the side of a press and after the operation . . . is finished, instead of being thrown on a pile on the floor, it will be loaded on another truck on the other side, ready to be carted to another machine for the next operation." The company processed about 18,000 tons of steel a year, and each piece of steel underwent an average of ten operations; it was picked up ten times and put down ten times, hence, under the old system "the plant handled 360,000 tons of steel." The new system eliminated most of the extra handling and "enabled the company to dispense with forty laborers."[8]

Other innovations to reduce handling costs included conveyor systems. The Gravity Carrier Company made conveyor systems that relied on gravity and made it possible "to deliver incoming merchandise directly from the railroad cars to any point inside [the factory]." The National Acme Manufacturing Company installed a "combined gravity and power factory conveyor system" that handled products "from the time they leave the automatic machines until they are loaded in railroad cars for shipping." The 225-foot-long conveyor not only carried the "steel, iron, and brass screws, nuts, bolts, and various special products" but also cleaned, dried, and rustproofed the screws. The company claimed that the new system reduced handling to a minimum, eliminated excess labor, increased productivity, and reduced the cost of finished parts—an impressive list of benefits for one relatively simple change.[9]

By around 1915 electric cranes, monorails, and internal railways aided the handling of heavy materials in many factories. In 1882 the Yale and Towne Manufacturing Company of Stamford, Connecticut, had built one of the first traveling cranes in the United States, but it was moved by a worker operating a rope pulley, and its effectiveness could not compare with the later electric-powered cranes. Horace Arnold claimed that "the electric-driven traveling crane revolutionized shop construction for heavy work . . . it effected such savings in handling work as to force its installation in every new shop."[10]

Companies installed cranes to move heavy equipment and materials within one department. The most common was the traveling crane, which consisted of a heavy beam extended across the width of the shop. The ends of the beams rested on wheels that traveled the length of the floor on rails. The hoist could be moved across the width of the beam, enabling the crane to lift materials from almost anywhere on the floor. Though used in many companies at the end of the nineteenth century, cranes became a standard feature in factories only in the first two decades of the twentieth. Cranes were used in a straightforward way—to lift heavy materials or parts or machines. When the Washington, Baltimore, and Annapolis Railway built a new repair shop in 1909, it installed a traveling crane as soon as possible. It was initially used in setting up the new shop machinery, and later it lifted railroad cars for repair as well as moving heavy running gear parts.[11]

Monorails allowed more flexible handling than the cranes. Not confined to one straight track like the crane, a monorail could cover an entire plant, go any distance, and travel around curves. Many factories chose it because it required a less expensive structure than the traveling crane even though it could not lift the same weight. Companies that installed monorails and other handling systems enjoyed savings as great as 65 percent.[12]

For many companies, moving parts and materials between plant buildings was as important as internal handling. In 1917 the Minneapolis Steel and Machinery Company used forty-seven men and fourteen wagons and teams of horses to haul loads between the plant buildings. In 1919 the company installed an "inter-shop haulage" system using ordinary industrial trucks with trailers. The change reduced the number of men used for handling to fourteen: superintendent, chief dispatcher, assistant chief dispatcher, five division dispatchers, four tractor drivers, and two men to operate the "jitney" service. It also increased the amount of steel moved by 50–100 percent. Despite those savings, in describing the significance of the new system the superintendent first pointed out that using machines instead of men and horses improved predictability: the "main savings isn't due to the smaller number of men necessary . . . it is because the right material goes to the correct place and gets there on time." The cranes and other heavy handling equipment, along with the increasingly heavy machine tools, strained the older factories. "A shop floor full of machine tools and covered by a traveling crane overhead demands a [type of] building . . . not known in this or any other country until late years."[13]

New Building Technology

Developments external to manufacturing made innovations in the factory building possible. The most important was reinforced concrete construction, which, along with structural steel, was used in industrial buildings beginning in the late nineteenth century. Before the introduction of reinforced concrete, all factory buildings were built of wood or masonry and were, for the most part, restricted to the style of the mill building. The new building material enabled engineers to build the more efficient factories they had envisioned.[14]

Reinforced, or structural, concrete was stronger than other types of construction material, including plain concrete. Reinforcing is accomplished by implanting steel bars or rods; this strengthens the concrete because it combines the elasticity of steel with the compression strength of concrete. Plain concrete crumbles when it is subjected to bending

Forming concrete slabs at Harbison-Walker Refractories, Wylam, Ala., 1909–1910. Cheap labor displaced the skilled carpenters who previously had built the framing for poured concrete, revolutionizing factory construction. Robert Cummings, the industrial engineer who designed the refractory, oversees work. Robert Cummings Papers, Smithsonian Institution.

forces such as those of a factory floor loaded with heavy equipment. Steel reinforcing bars give the concrete the tensile strength to withstand that kind of stress.[15]

Reinforced concrete changed the design and construction of factories. Its strength allowed a building to be supported on its internal columns, relieving the outside walls of bearing any of the building's weight. This single advancement opened the way for changes that ultimately transformed the factory.

Reinforced concrete accomplished four improvements in factory construction: it reduced floor vibration from machines, making multiple stories more feasible; it required fewer interior columns than older mill construction, opening up space on the shop floor; the strength of the concrete outer walls meant both that the window areas could be much larger, opening the way for the window-walls of the daylight factory, and that the building could easily be made much larger than the mill building. These changes eliminated the major deficiencies of traditional mill construction.[16]

In addition to providing added strength, concrete almost eliminated fires in factories. The owners of preconcrete factories lived in constant fear of fire that would wipe out their entire facility as well as the neighborhood, or more. The concern was so great that most nineteenth-century changes in factory construction had been attempts to decrease the danger of fire. Slow-burn mill construction—the name revealing the hopes of its inventor—replaced the lighter and more easily combustible construction styles of the early nineteenth century. Mill owners used heavy beams that were slower to burn, built flat instead of pitched roofs to eliminate the dead space of roof rafters where fires often started, placed stairs in outside towers, and installed sprinkler systems.[17]

Soon after 1910 factory owners realized that concrete buildings were so fireproof that many stopped paying for fire insurance. Concrete's success at preventing fires is reflected in insurance rates of the period (see table below).

Insurance Rates per $100 of Value

Wood Mill Construction Wood Sides		Wood Mill Construction Brick Sides		Concrete	
Bldg.	Contents	Bldg.	Contents	Bldg.	Contents
.75–1.50	1.00–2.00	.20–.75	.60–1.00	.10–.40	.39–.70

SOURCE: Henry G. Tyrrell, *Engineering of Shops and Factories* (New York: McGraw-Hill, 1912), p. 54.

At the turn of the century, many engineers and builders patented their own designs for reinforcing rods—round rods, square rods, rods twisted in dozens of different ways, looped rods, and hooked rods. One important discovery was that the only way to strengthen concrete was to suspend the rods in the middle of the concrete mass; rods simply pushed in the concrete in any position did not improve its strength.[18]

The concrete buildings of the late nineteenth century were monolithic in construction—unreinforced concrete poured into forms that shaped floors, walls, and roof. These buildings were expensive and often difficult to build. Problems arose first because the unreinforced concrete had to be used in massive amounts to assure its strength. Only relatively small buildings could be constructed with the monolithic technique because

Acme Screw Machine Company machine shop, circa 1913. Before the introduction of single-drive motors, similar jungles of belting beset all machine
shops. Spooner and Wells Photo Collection, Neg. 0-3295, Ford Motor Company Archives, Dearborn, Mich.

the concrete had to be poured in complete units. Half a column, or half a
floor, could not be poured one day and half the next because wet concrete does not adhere well to dry.

Monolithic concrete buildings were also expensive because the forms
had to be built by skilled carpenters. When the concrete had dried after a
few days, the forms were torn down and discarded. The forms that
created the outline of the entire building proved the most expensive and
time-consuming component. "Forty to sixty percent of the cost of concrete work is right there in the forms," reported engineer Robert Cummings in 1910, "so that if you eliminate the forms you are getting a more
economical form of construction."[19]

The expense of concrete construction limited its use even after builders and architects understood how to use reinforcing rods. The industrial
architect Albert Kahn explained that reinforced concrete "had been in

use for some time abroad where labor costs were lower, but adopted here only hesitantly because of its greater cost and the danger connected with its use."[20] By the early years of the twentieth century, concrete builders developed a new technique that reduced both cost and construction time. The new method, slab construction, used unskilled laborers to pour slabs measuring approximately four by six feet. Laborers built the simple forms and poured the concrete slabs on the ground, and the slabs were then assembled much like bricks. The slab method reduced cost because it used inexpensive labor and reused forms. It was also faster because forms did not have to be constructed and torn down; many slabs could be poured in advance, ready for use when needed. Concrete construction, and especially the slab method, led to larger buildings and changed the nature of construction work.

Electrification

Important innovations also resulted from factory electrification. Electricity in the factory meant two things: electric lighting and electric motors. Electric lighting, installed before electric power, improved visibility and safety and generally improved the working environment. It also raised productivity because it made twenty-four-hour production more feasible. As important as it was, though, electric light did not change thinking about how to organize or run a factory; electricity's major impact on plant organization lay in the new possibilities for shop floor layout inherent in the use of the electric motor.

Before electricity was installed to run factory machines, power distribution dictated the layout of the shop floor. All nineteenth-century factories relied on a central power source—a water wheel early in the century or a steam engine later. The water wheel or steam engine turned a vertical shaft that extended upward through each floor and connected to a cumbersome system of shafting, gears, and belts to deliver energy from the power source to individual machines. On each floor a line shaft, attached to the ceiling, connected to the main shaft and extended the length of the floor; it turned as the main shaft turned. Belts and gears, attached to the line shaft, powered individual machines.[21] Machines had to be placed parallel to the shafting, and the shaft turned constantly.

This system had inherent problems. The multiple transmissions meant lost energy, and one part of the factory could not be used without all of the power being engaged. The system could also be dangerous because the belts from line shaft to machine often broke, injuring workmen. The belts interfered with movement through the factory, blocked light, and needed constant maintenance. Perhaps the most significant limitation to production was that the layout of the shop floor was restricted to parallel placement of machines. The system made expansion difficult because building size was limited according to the power system. If expansion was not anticipated in the original plan and construction, a new power system had to be installed to accommodate new factory space.

In the 1880s and 1890s, engineers and industrialists debated electrification's potential effects on industry. Some argued that electricity would result in savings in energy and capital. Others believed that while electricity had many advantages, it would not result in a significant cost reduction because most establishments spent no more than 3 percent of their operating costs on energy production. Industries such as textiles used a steady amount of power throughout the day, and the power load remained constant through the work week. For those industries the older power sources continued to be sufficient, and they put off the expensive conversion to electricity.[22]

The first factories to use electric power simply installed generators to replace the steam engine as a central power source; the new engines turned the shafts previously turned by water or steam. This changed nothing about the way the plant worked. Power transmission to machines and machine placement remained unchanged. By the end of the nineteenth century, however, some engineers advocated group drive to replace central power sources. Industries that required intermittent use of machinery would profit by installing electric motors to drive small groups of machines. Belting and shafting still transferred power from the motor to the machines, but one group of machines could be operated without engaging the entire factory. Group drive allowed changes in the shop floor layout because the line shaft no longer dictated placement.

The most important change came with unit drive machinery—one electric motor attached to or built into one machine. Unit drive allowed independent operation of machines using no belts and no shafting. It increased range of machine speed, efficiency of power use, and flexibility in machine control.[23] Unit drive systems allowed the industrial engineer new freedom to experiment with shop floor layout. Without the shafts the engineer moved away from the parallel rows of machines; new arrangements improved production flow, made better use of floor space, and increased efficiency. All of this helped the industrial engineer move toward his goal of increased production, which in many cases rose by 20 or 30 percent with the new arrangements.[24]

Other improvements in the factory resulted from electrification. Some engineers hailed the improved appearance of the shop floor and claimed improved morale. Removal of the cumbersome belting resulted in better light distribution and cleaner factories. The new power system reduced fire risk by eliminating the openings in floors through which the old shafting passed. Electricity opened vast possibilities for the industrial engineer, and many industries eagerly exploited those opportunities.

The automobile industry provides one of the best illustrations of the important changes introduced in the early twentieth century. It is a good example because the industry, and especially the Ford Motor Company, experimented with the new ideas about organizing factories within a short period of time; Ford, for example, went from a small machine shop in 1904 to a major experiment in planning a rational factory in 1914. That the industry grew so rapidly at the same time as industrial engineers were developing their practical ideas about the factory as machine was fortuitous; the coincidence enabled auto companies to streamline their production techniques as they built one of the largest industries in the world.

Early Automobile Factories

The early automobile industry, from 1900 to about 1920, typifies the kind of changes that engineers made in early-twentieth-century factories. It is useful to consider the early history of the industry—the years when auto companies were nondistinctive elements of the industrial landscape—before moving to a discussion of the revolutionary auto factories.

The young industry, having few requirements, started out in small machine shops. Even the most successful early companies such as the Olds Motor Works, which built fairly large factories, had no special requirements for its factory buildings. The builders of auto factories during the first few years of the century demonstrated little concern with innovations being introduced in other types of factory. But as auto companies experienced their unprecedented growth and revolutionized production in the second decade of the century, they began to borrow from developments in other industries and then went farther, to build rational factories.

The fledgling auto industry was concentrated in and around Detroit at

the end of the nineteenth century. It attracted many inventors and entre-
preneurs, and hundreds of men tried their hand at starting small auto
companies. Only a few succeeded. A 1904 Michigan Bureau of Labor
Statistics report acknowledged only three Detroit auto companies—Olds
Motor Works, Buick, and Packard—suggesting that the later giant, the
Ford Motor Company, was not yet a contender for an important place in
Michigan's industrial community.[25]

Ransom Olds, the man behind one of the most successful early auto
companies in the United States, began production of the Oldsmobile in
1899 with a capital stock of $500,000. Olds was the first to bring the
automobile into quantity production and the first to build a factory
described as being "designed and laid out for the manufacture of the
motor car." In Olds's first factory, workers could build eighteen cars per
day. Increased demand encouraged him to add to his first plant, raising
daily production capacity to fifty cars; he also built another plant of
similar size in Lansing in 1902. By 1903 building additions increased floor
space to 570,000 square feet for the two plants.[26] Olds's factory may have
been the first built specifically to manufacture automobiles, but there is
no evidence that it differed from plants in other industries. The buildings were typical masonry mill buildings, arranged to create a hollow
square. The few interior photographs that exist show wooden floors and
columns and windows of the size required by load-bearing construction.

In 1903 the Packard Motor Car Company set a new course for the construction of automobile factories when its president, Henry Joy, hired a
young architect to design and supervise the building of a new plant.
From 1903 to 1905, Albert Kahn designed and oversaw construction of
nine buildings for Packard, all of standard mill design and organized
much like other factory complexes. In 1905, as he built the company's
tenth building, Albert and his engineer brother, Julius, began their experiment with reinforced concrete. Packard Number Ten and its architect attracted considerable attention from the industrial community. To
Detroit industrialists the building represented a significant advance in
factory construction. In his use of reinforced concrete, Kahn played a
role in the revolution in industrial building by helping to eliminate the
deficiencies in mill construction.[27]

Most automobile shops of the early years of the century could not
compare to the Olds and Packard factories. According to the 1904 Michigan Bureau of Labor Statistics report "in general a large part of the

modern factory for the manufacture of automobiles does not differ materially from that which will be found in any well-equipped machine shop adapted to produce 'parts' in large quantities." In his 1926 reflections on the auto industry, Fred Colvin, a prominent industrial journalist, reminded his readers that the early auto industry "was not particularly important. The average automobile shop . . . was building automobiles in a small way and by ordinary machine shop methods."[28]

Most of the early auto companies designed and assembled cars, buying components from small manufacturers (Olds, Buick, and later, Ford were exceptions in manufacturing many of their own components). The job of assembling a car in 1903 required the skill of an experienced hand and the use of standard machine tools. Every piece had to be cut, filed, and fitted because the industry had not yet achieved the production of standardized, interchangeable parts. The work was done by small groups of skilled machinists working together to build one car at a time. The construction of an automobile depended on human skill rather than the skill of any special-purpose machine, even though such machines were commonly used in other industries by that time.[29] The shop floor was divided among small working groups, each assembling a single car. Management was minimal; in most companies both owner and foreman were skilled machinists rather than school-educated engineers or managers, a situation that created a collegial rather than hierarchical relationship.

Though the early auto workers left little behind to tell us what they thought of their jobs, we can imagine that the work—a respected trade with a small number of fellow workers in a new and exciting industry— was agreeable to a skilled mechanic. The work was anything but routine, and it presented challenges and paid good wages to workers with the skill to do the job. Auto industry wages were among the highest in the country: in 1914 they ranked seventh in average annual terms, by 1919 they were fifth, and by 1925, first.[30]

The first factory of the Ford Motor Company was like other small auto shops. A small, inconspicuous, wood-frame building on Mack Avenue in Detroit, it consisted of only one room and was later expanded to two by the addition of a second story. In 1903 one of Ford's first investors wrote: "The building is a dandy. I went all through it today. It is large, light, and airy, about 250 feet long by 50 feet wide, fitted up with machinery necessary to assembling the parts, and all ready for business." The shop's machine inventory consisted of two lathes, two drill presses, one milling

Ford's first factory on Mack Avenue. Built in 1903, it met the simple needs of the company during its first years. Neg. 0-4135, Ford Motor Company Archives, Dearborn, Mich.

machine, one hand saw, one grinding wheel, and one forge; it contained no specialized tools and employed a foreman along with "ten or a dozen" men.[31]

The design of auto factories changed little during the first few years of the century, a time when the production process remained stable and made few demands on the factory building. But as the demand for autos grew, pressures mounted for increased speed and volume of production, and there was a heightened awareness of the need to develop appropriate plant design. In 1904, Ford's success with his Models C, F, and B made him think about moving to larger facilities. When the company began construction of a second factory on Piquette Avenue in Detroit, Ford and his associates knew that they needed a shop significantly different from the Mack Avenue one. It provided better organization and more space for growing production but did not offer a new model for production or factory design.

The three-story building measured 402 by 52 feet on a lot about four times as large as the building. Though Ford's new plant almost tripled the company's space, its 63,000 square feet hardly compared to the 570,000 square feet of the Olds plant in 1903. The dimensions of Ford's Piquette Avenue factory were typical of most nineteenth- and early-twentieth-century factories, which depended on the sun to help light the shop floor; no industries relied solely on electric lighting yet. The standard mill building was still considered the best design for catching daylight and distributing power.[32]

The Ford Motor Company and the other automakers could produce automobiles in plants like the one on Piquette Avenue because the process of building an automobile differed little from the way any other heavy goods, such as carriages, wagons, or machinery, were built. They were made the only way anyone knew how to make cars: one at a time by workers in small groups, filing and fitting parts.

Though the Piquette Avenue factory resembled the company's first shop in its production operations, it differed because it contained more than just workshops. A visitor would enter the first floor through a proper lobby with a receptionist and might have been directed to one of the private offices of the company's officers; such formality was unknown in the Mack Avenue shop's two rooms. The rest of Piquette's first floor contained the general offices for clerks and stenographers as well as storage areas and a machine shop with the heaviest machines—the cylinder, crank case, and crankshaft departments. The first floor would have seemed dark because of windows slightly smaller than those on the top two floors and the stacks of supplies around the perimeter. The second floor looked and sounded different from the first; larger and lighter, it housed a variety of operations from assembly stations and machine shop to offices and Henry Ford's private experimental room. The third floor looked much like the second; it held pattern rooms, storage, and the final assembly area where twelve to fifteen groups of men did the final work of putting the car together.[33] This organization could easily be changed if models or production methods changed. Nothing in this small factory was heavy or immovable.

At the end of 1905, when Ford set up the separate Ford Manufacturing Company on Bellevue Avenue to make engines and rear axles, he followed the examples of Olds and Buick in establishing a major in-house manufacturing operation. It was there that Ford and his foremen began to rethink work organization.[34] With the help of Walter Flanders, a New England machinist, the company experimented with the manufacture of interchangeable parts. To ensure interchangeability, parts were machined with greater precision—each one had to be exactly the same. Fixtures, jigs, and gauges added to machine tools assured precise machining of each piece and also allowed less skilled men to do the job.[35]

To speed production, machines were arranged sequentially; rather than placing machine tools in the traditional pattern with groups of like machines together, the new arrangement placed machines in the order in which they were used, reducing time and labor costs of transporting

Above, Ford's Piquette Avenue factory, built in 1904 and in service to 1910. Though much more substantial than the simple, garagelike Mack Avenue plant, this building incorporated no distinctive or function-related features; *below,* inside, workers employed pre–assembly line production techniques. Negs. 0-658 and 833.39894, Ford Motor Company Archives, Dearborn, Mich.

THIRD FLOOR

SECOND FLOOR

FIRST FLOOR

FIRST FLOOR: *1*, private office; *2*, reception room; *3*, lobby; *4*, office of J. Couzens; *5*, vault; *6*, bookkeeping; *7*, sales office; *8*, clerks and stenographers; *9*, men's toilet; *10*, ladies' toilet; *11*, office; *12*, auditing; *13*, lobby; *14*, time clock; *15*, time and employment office; *16*, factory men's toilet; *17*, final test area for model K's; *18*, storage; *19*, wash rack for model K's; *20*, audit storage; *21*, first floor machine shop (pistons, bolts, nuts, etc.); *21.1*, Porter and Johnson's; *21.2*, Acme Automatics; *22–23*, storage; *24–25*, shipping; *26*, storage battery assembly.

SECOND FLOOR: *1*, office of Henry Ford; *2*, vault; *3*, office of C. H. Wills; *4*, office of W. Flanders; *5*, office of Walborn; *6*, office of P. E. Martin; *7*, secretary's desk; *8*, experimental tool room; *9*, experimental room; *10*, parts to be machined; *11*, factory men's toilet; *12*, model K assembly; *12.1*, desk of assembly super; *13*, storage; *14*, machine shop for models N, R, and S; *15*, storage; *16*, model K magneto room; *17*, body painting and trimming.

THIRD FLOOR: *1*, microphotographic room for vanadium analysis; *2*, drafting room; *3*, wood pattern room; *4*, metal pattern room; *5*, men's toilet; *6*, models N, R, and S assembly; *6.1–6.3*, stand up desks; *7*, storage; *8*, wiring assembly for models N, R, and S; *9*, body assembly for N, R, and S; *10*, unfinished body storage; *11*, storage; *12*, Joe Galamb's secret room.

Above, Author's drawing of floor plan from a 1907 plant layout and notes made by H. L. Maher in 1953 based on his memory of working in the factory; see Piquette drawings, X3358-60, Blueprints and Drawings, Ford Motor Company Archives, Dearborn, Mich.

partially completed operations between machines. "Before that," noted one of the company's engineers, "we had all lathes in one place, all drill presses in another and then we moved the materials around from place to place. We then worked out the idea of sequence to avoid carrying the material from one machine to another." No longer would men work in small groups, each doing whatever needed to be done; instead, for final assembly they would walk along a row of stationary cars performing the same operation on one car after another. Ransom Olds had already experimented with this method of final assembly; as early as 1904 he had assigned the assemblers to a specific task, arranged the parts they needed in large bins at their work stations, and had the cars move from station to station on wooden platforms.[36] Ford's experiment was soon expanded to the assembly area of the main Piquette Avenue building, where Ford and his engineers initiated changes that would lead the company to its assembly line production in just a few years.

When transportation by horse cart between the manufacturing plant on Bellevue Avenue and the assembly plant on Piquette Avenue, a distance of about four miles, proved to be too slow, the company purchased additional buildings next to the plant on Piquette Avenue and in 1907 moved engine production closer to assembly operations.[37] Despite the expansion, Ford was constrained by lack of space to try out his ideas. That constraint, combined with a surprising increase in demand for his cars, led Ford and his engineers to their experiments with plant design which helped to create the momentum that changed the American factory.

The Crystal Palace: The Ford Motor Company's Highland Park Plant, 1910–1914

BY 1910 NEW ASSUMPTIONS ABOUT PRODUCTION DICTATED IM-portant changes in the planning and design of factories. It was a busy year as many automobile companies expanded their factories or moved to new ones. In April the Pierce-Arrow Motor Car Company of Buffalo, New York, expanded, and the Clark Power Wagon Company of Lansing, Michigan, built its first factory. In July the Dodge brothers built a new plant in Detroit, and the Dayton Motor Car Company added new build-ings to its plant.[1] The move that attracted the most attention was the Ford Motor Company's relocation to its new plant just outside the De-troit city limits in Highland Park. Ford's new plant represented the be-ginning of a new era in the design of automobile factories, signaling a significant step in the development of the rational factory—a predictable and obedient machine. In theory the rational factory would not need to depend on the welfare programs used by some manufacturers to moti-vate workers but would rely solely on industrial engineering practice.

The auto industry, and the Ford Motor Company in particular, pro-vides the focus of the following two-decades-long case study in the de-velopment of the rational factory. In a discussion of the rational factory, the auto industry is the most significant of twentieth-century industries because it was a major source of technological and managerial innovation. Instrumental in the final development of the rational factory, the auto industry created an exacting industrial environment that combined more different kinds of operations on a larger scale than any other industry. Its demand for a well-designed factory went beyond its earlier requirements for a structure strong enough to hold the heavy machinery and with enough windows to provide adequate light. The growing enterprise required a building able to hold machines that seemed to grow heavier each year, but its engineers also intended the building's design and layout to facilitate both production flow and the management of workers. Indeed, the combination of demands—careful management of workers, efficient handling of materials, and increased speed of production—became the test that determined whether a company would survive in a tough business climate.

Engineers changed the way they designed factories in the early twentieth century because the factory building, like other industrial technology, had become an essential element in increasing productivity. The new buildings had become a tool of management, and plant design became one of the techniques used by growing companies to increase speed and volume of production and, most importantly, to assure a steady flow of production.[2] This chapter and chapters 6 and 7 examine that development at the Ford Motor Company.

The Daylight Factory

The first feature that impressed most observers about Ford's new Highland Park factory was the expanse of windows. Highland Park has been almost universally praised by architectural and industrial historians of the United States as the first "daylight" factory. Reyner Banham has shown, however, that several "daylight factories" in Buffalo, New York, predate Highland Park.[3] But why were the large windows deemed so important when, by the time of their introduction in 1910, electricity could have provided lighting adequate for industrial production? The large windows did not always provide increased ventilation, as was often claimed, in many factories only a few panes in any section actually opened. The large areas of windows were expensive to replace when broken; they also had to be kept clean if their benefits were to be enjoyed, and keeping them clean was no small task.

One answer lies in the industrial welfare movement's efforts to rationalize the human machine: to attract quality workers and reduce labor turnover, social engineers and welfare secretaries introduced design features that improved the quality of the work environment. Problems with labor turnover were especially acute in Detroit. Auto companies competed for skilled workers, who frequently changed jobs; by 1913 the Ford Motor Company had a turnover rate of 370 percent.[4] One way to keep good workers was to build comfortable factories. Factory beautification was part of the industrial welfare movement's focus on improving worker morale through social and environmental improvements such as libraries, classes, gardens and better factories. The daylight factory was considered a revolution in improving working conditions by opening up the dark interiors of mill buildings.

Some of the interest in the daylight factory, no doubt, came from ideas inherited from the larger social reform movements of the late nineteenth and early twentieth centuries. Reformers heralded the importance of abundant sunlight and ventilation for their purifying effects. The long-standing belief in sunlight's sanitizing qualities persisted even when turn-of-the-century scientists found more effective measures to improve the health conditions of the lower classes.[5]

Though the widespread popularity of the daylight factory was probably a result of the industrial welfare movement, its very existence was a product of architectural and technological innovation. The German architect Peter Behrens had already used windows in a unique way in his AEG Turbine factory in Berlin, but it was not yet a daylight factory. Other architects were also experimenting with new factory designs in the northeast United States. So even though Albert Kahn, the architect for the automobile industry, received the greatest attention for his early daylight factories, he shared interest in the new type of factory with many others. Daylight factories became practical only with the introduction of reinforced concrete.

Highland Park may have been the first factory in the United States to use the kind of steel window frames that allowed maximum window space, but it was not the first attempt to open up the factory to sunlight. A number of reinforced concrete factories had already been built.[6] Undoubtedly many daylight factories existed which have escaped the historian's attention, such as the reinforced concrete machine shop built at McKeesport, Pennsylvania, in 1904 by Robert Cummings, a structural engineer. Concerned about light in the plant, he designed the factory with abundant windows, which except for larger frames of wood, admit

Robert Cummings's building for Taylor-Wilson Manufacturing Company, McKees Rocks, Pa., 1905, illustrating an early daylight factory. Robert Cummings Papers, Smithsonian Institution.

as much light as Highland Park's windows. Cummings's concern for adequate lighting for the workers is reflected in correspondence with the firm, though the company's lack of interest is inconsistent with the general mood regarding sunlit factories.

> Firm to Cummings, June 21, 1905
> We do not see our way clear to follow your suggestion in regard to white washing this building, as the light would be entirely too strong for the workmen.

> Cummings to Firm, June 26, 1905
> My suggestion as to white washing the interior of your building was based on supposition that too much light for the mechanics could not be obtained.[7]

Cummings's building could easily have escaped the attention of the Kahn brothers, but similar German factories could not have. German immigrants, Albert and Julius frequently returned to Germany and visited other European countries where well-lit factories could be found before Highland Park was built in 1910.[8] We can presume that one or both of the brothers visited some of those factories. Surviving Kahn records are scanty, so it is impossible to know if they introduced window walls in the spirit of architectural style or if they realized that opening up the factory to sunlight was important to the industry.

Though Albert Kahn has typically been credited with designing and

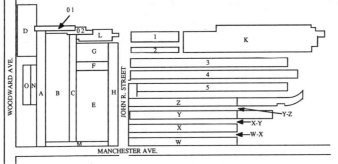

A, Four-story factory building; *B,* one-story machine; *C,* one-story craneway; *D,*
five-story power house; *E,* one-story machine shop; *F,* one-story craneway; *H,*
four-story factory building; *K,* foundry; *L,* gas producer and refrig. plant;
M, four-story factory building; *N,* garage; *O,* four-story administration building;
W, X, Y, Z, six-story factory building; *o1,* coal bunker and conveyor; *o2,* unloading dock; *1–2,* heat treatment building; *3–4,* forge; *5,* factory and storeroom.

Above, Highland Park's first structure and eventual layout. *Below,* The first
building (A), completed in early 1910, became known as the Crystal Palace
because of its unusual "window walls." Buildings on the left of John R. Street
constituted the Old Shop; buildings on the right made up the New Shop and
clearly demonstrated the company's new ideas about plant organization.
Photo courtesy of Albert Kahn Associates; author's drawing from plans in
Insurance Appraisal, 1919, Acc. 73, box 1, Ford Motor Company Archives,
Dearborn, Mich.

building Highland Park, his brother Julius clearly played an influential role. But in spite of the acclaim they have received, they are probably responsible for designing only the building's shell. Only vague references remain, but they suggest that Ford's engineers designed the layout of the factory buildings and turned the basic plans over to the Kahns to enclose.

When Kahn worked on the Highland Park plant, industry was not yet a desirable client to most architects in the United States. Kahn's interest in industrial buildings was probably a combination of serendipity and the influence of the noted German architect Peter Behrens. Though Kahn does not mention Behrens specifically, the Kahn papers contain numerous letters describing their tours of German architecture. Behrens's work was especially noteworthy because it had gained the attention of the art world and because of his unusual client—industry. The German architect's work could not have escaped the Kahn brothers. Behrens, famous as artistic advisor for Germany's AEG,[9] wrote about architecture's importance to industry long before his counterparts in other industrial countries. Behrens's work with AEG began in 1907 when he embarked on an extraordinary career as industrial designer, architect, and writer. With remarkable foresight, he wrote essays about industry, society, and architecture which were widely read in Europe. Behrens believed that through good design he could alleviate some of the social and personal devastation wrought by industry. He also believed that industrial efficiency grew out of good design.[10] Though Albert Kahn was undoubtedly influenced by Behrens, the imagination and idealism of Behrens is hard to find in Kahn's work or writings.

Kahn's first factory commission came from Henry B. Joy, whose house Kahn had designed. Impressed with the architect's work, Joy hired him to build the new Packard Motor Car factories. In his own words, Kahn undertook the job "with practically no experience in factory building." He admitted that Julius, his brother the structural engineer, played a very important role in the eventual success of the Packard plant.[11]

George Thompson, one of Ford's engineers, later described Kahn's special character that enabled him to work so well with industry. "Mr. Kahn was a wonderful architect but he didn't know anything about [the industry]. He was extremely quick in grasping his client's needs. He would be sitting there sketching almost as fast as anybody could talk as to what it was to be, and would be back the next morning with sketches."[12] Kahn's speed undoubtedly appealed to Ford!

Ford's New Factory

In 1908 the Ford Motor Company introduced the Model T. It became the single most popular model ever produced in the United States and helped to persuade a highly skeptical population that the motor car was a good idea. It was strong, durable, easy to drive, easy to fix, and so successful that for eighteen years it was the only car Ford made. During that time it became the object of much affection in the country, with songs and stories written about it. The T's fame assured Henry Ford's, and its success made him one of the richest men in the world.

Much of the T's success came from its low price, which ranged from $850 in 1908 to $290 in 1926, with a low of $260 in 1924.[13] From the beginning, it outstripped all competitors in sales, and the small firm found demand for the T impossible to fill. To manufacture the low-cost automobiles, Ford established a minute division of labor using highly specialized machines that required a minimum of skill to operate. Eventually, the work would be connected by a moving assembly line that automatically set the time allowed for any operation.

By 1909 the Ford Motor Company was passing all competitors; business was booming with Model T sales, and Ford began to think about his third plant. As Henry Ford thought about more space, he envisioned more than a simple expansion of the existing facility; he imagined a different kind of factory. Ford and his engineers were trying to combine all their innovations and all they knew about production into one big plan. They decided on an internal factory arrangement quite different from that at Piquette Avenue, an arrangement that would facilitate expansion *and* the new production process that was just beginning to take shape. Ford turned to the architect who seemed to be the most inventive and hired Albert Kahn. At the first meeting between the two, the automaker told the architect: "I want the whole thing under one roof. If you can design it the way I want it, say so and do it."[14] Kahn did, and continued building for Ford and other automakers.

In January 1910 the Ford Motor Company moved into a group of nine buildings in Highland Park, Michigan, just outside the Detroit city limits. It was a group of buildings that served as the beginning of a larger industrial complex. The company's innovations in technology, management, and marketing, not to mention its eccentric president, constantly drew public attention to the Highland Park plant. The industrial and

architectural worlds took special notice of the new buildings. They recognized immediately that this was no run-of-the-mill factory; its manufacturing space alone, not counting the office space and powerhouse,
was double the size of the Oldsmobile plant (then the largest in the Detroit), but most important was the internal organization, which changed
constantly as Ford and his engineers explored ways to make production
more efficient.

The most obvious structural feature of the plant, which even the
casual observer would notice, was the expanse of windows which replaced the brick walls of the traditional factory. The window walls inspired observers to call the building the Crystal Palace, a reference to
London's famous glass hall built by Joseph Paxton for the Great Exhibi-
tion of 1851.

Most of Highland Park's buildings were of concrete slab construction
with concrete-covered steel girder beams and a brick facing. The concrete
held reinforcing rods designed by Julius Kahn.[15] Building A, the main
factory building, was four stories high, 860 feet long, and 75 feet wide; it
held assembly operations, stockrooms, tool cribs, painting booths, and
other light operations. The proportions of Building A, and the other
four-story buildings, were those of the traditional mill building, but there
the similarity stopped. The interior spaces were those of a modern,
reinforced-concrete factory, with large, uninterrupted floor space. In
order to leave open as much space as possible on the shop floor, four
utility towers, containing the elevators, stairs, and toilets, were erected
on the outside of the building.[16] Two additional four-story buildings,
which held assembly operations, joined with Building A to form a hollow
square. Though the concrete construction made the building stronger
than most factory buildings, according to an engineering appraisal years
later, the construction of the four-story buildings was inadequate for
heavy manufacturing, having capacity for only about fifty pounds per
square foot, a load limit suitable only for assembly and storage.[17]

The new plant also housed a large and comprehensive machine shop
situated in the single-story space created in the center of the hollow
square arrangement. The machine shop was the heart of the factory; a
jungle of leather belting and gears, it shaped the essential parts of the
Model T. Except for the abundant light admitted through the saw-tooth
roof with its north-facing windows, the closely placed machines and the
power belts would have made the machine shop a dark and gloomy part

of the factory. An English invention, the saw-tooth roof took its name from its jagged profile, which resembled the tooth edge of a saw. The special roof style was common on industrial buildings because it provided evenly distributed, well-diffused and nonglaring northern light that eliminated sharp shadows or contrasts. Machinists preferred the diffused light for their precision work.[18]

Ford's move to Highland Park signified the beginning of major changes in personnel, new spatial relationships within the factory, and new social patterns as well. For Ford himself, the move meant that he was less involved in factory life than he had been earlier. He located his office in the new administration building, built at the front of the plant complex and surrounded on three sides by an attractive lawn. The most significant feature of this building was not the elaborately furnished lobby with high, carved ceilings, wide, curving staircase, and marble floors, or the fact that it housed the only dining facility in the plant, but that it was a freestanding building, separate from the factory buildings. In Ford's first two factories, workers and managers had enjoyed close working arrangements that reinforced the lack of rigid hierarchy. But by 1910 the industry had changed; in the new plant, the highest level of management was housed in a fancy, segregated building, a fact that reflected a new relationship between management and worker.

Along with the administration building, the plant's powerhouse became the company's public face; both were showplaces built at considerable expense. Ford took great pride in the powerhouse, wanting it visible and insisting that it be kept spotless. He believed that the public was as fascinated with the huge generators as he was, and he had the building walled in plate glass. He also believed that it would be a great advertisement of the company's power and modernity. More than glass made the powerhouse a showcase. It contained the finest fittings available: brass and copper fixtures, chains, and rosettes, mahogany handrails, and "all workmanship first-class in every particular." Atop the powerhouse hung the huge FORD sign between the five stacks. William Vernor, the mechanical engineer for the powerhouse, explained years later that only two stacks were needed, but Henry Ford wanted the sign and insisted it be hung between powerhouse stacks, so there had to be five.[19]

The company claimed that the two Ford buildings were unusual in their elegance, but in fact other companies made similar efforts to attract attention to their plants. Earlier factories also boasted "the handsomest

Highland Park's powerhouse, with an extra smokestack to accommodate the Ford name. Large, displaylike windows invited pedestrians on Woodward Avenue to stop and admire the powerful generators and gleaming brass fittings. Ford insisted that the powerhouse—as an advertisement for the firm—remain spotless. Neg. P.O. 3410, Ford Motor Company Archives, Dearborn, Mich.

engine room in the country," and many plants displayed administration buildings of fine architecture.[20]

The role of the powerhouse as a company showplace was also facilitated by its proximity to the street and sidewalk—it was accessible to pedestrians, who could admire the gleaming generators as they passed. When Ford built the Highland Park plant, it was just outside the official city line in an undeveloped area, but it didn't take long before it became a dense and thriving urban neighborhood with the plant at its center. This was an urban factory, part of urban life. Sidewalks led right up to the plant; it was not surrounded by high fences and the kind of industrial no man's land that would characterize later plants.

This accessibility also meant that workers walked to work and stepped from the sidewalk through the plant gate. Such access proved convenient for labor organizers; when International Workers of the World (IWW) members began to organize the company in 1913, Ford workers listened to union speeches outside the main entrance during their lunch break. When police dispersed the crowd at the gates, the unionists merely relocated down the block in an empty lot. This easy access to the work force would be impossible in later plants, where barriers to public access were deliberately constructed.[21]

Production

Industrialists and engineers considered the Highland Park plant special for reasons other than its reinforced concrete and acres of glass. It housed the first experiments with the moving assembly line and Ford's particular style of mass production. The production innovations that Ford engineers introduced in the Highland Park plant have been so important to understanding twentieth-century mass production that many historians have described them. The excellent descriptions available elsewhere allow me simply to summarize the innovations on the shop floor.[22]

The company's production system, only later called Fordism, developed around deliberate decisions: produce a single model, inexpensive but good, in large quantities, with a small margin of profit. The system depended on standardization, mechanization, speed, efficiency, and careful control of production and workers. As production began in the new plant, engineers focused on division of labor and specialization of work. In 1910 jobs that had been divided at Piquette were further divided, and workers were assigned to the production or assembly of a particular section of the car. At the engine assembly stations, for example, men worked at open workbenches with parts bins placed in the middle of the benches, each man assembling one engine. Dashboard assembly stands, placed with just enough room to allow hand trucks to move between them, held the dashboard unit while the worker attached the parts from the bins at the bottom of the stand. Similar operations characterized magneto, rear axle, and radiator assembly. Final assembly remained static in 1910—the car frame rested on wooden assembly horses or stands while assembly teams moved down the row of chassis, each gang performing a specific task or series of tasks.[23]

The machine shop contained sophisticated, highly specialized machines designed to do one operation over and over; they allowed the work of skilled machinists to be imitated by less skilled workers. No longer were all of the machine shop workers necessarily experienced machinists. Ford later declared that he would rather have operators with no experience at all; he wanted workers who had "nothing to unlearn" and who would work just as they were told.[24] Reducing reliance on human know-how was part of the intent of the rational factory.

The new factory meant new work environments. The single operation assigned to a worker, whether in the machine shop or assembly, meant that he repeated one task for the entire day. He was not allowed to move around the shop floor, much less the rest of the factory. All parts and supplies were delivered to him at his work station. In contrast to the Piquette Avenue days, when workers had enjoyed control over how they performed a task, Highland Park workers had little discretion over their time or work procedures, and the little bit that remained in 1910 would soon be gone.[25] Ford engineers realized that, to achieve higher production volume, they had to change the way workers worked. In the Highland Park plant they began the process of limiting the workers' freedom of movement around the factory and discretion over tasks and timing.[26] The first step was to limit acceptable travel away from work stations. As auto production was increasingly rationalized, management would try to control as much of the worker's movement as possible.

The new plant contributed to Ford's considerable success—fiscal year 1910–11 saw production more than double, and it doubled again in each of the following two years. At the same time the number of hours it took to build a car decreased each year. These successes were like an addiction to Ford and his engineers, and they constantly searched for new ways to increase production speed as they reduced costs; almost every month brought new innovations that put the company a step closer to a rational, automatic factory (see tables below).

Production Rates

1909–10	14,000	1912–13	168,000	1915–16	534,000
1910–11	35,000	1913–14	248,000	1916–17	785,000
1911–12	78,000	1914–15	308,000	1917–18	707,000

SOURCE: Henry Ford, *My Life and Work* (Garden City, N.Y.: Doubleday, Page, 1922), p. 145. Numbers of automobiles rounded to nearest thousand.

Average Man-Hours per Car Built for Each Available Year

1912	1,260	1915	533	1921	322
1913	966	1919	415	1922	273
1914	617	1920	396	1923	228

SOURCE: Martin La Fever, "Workers, Machinery, and Production in the Automobile Industry," *Monthly Labor Review* 19 (Oct. 1924): 738.

The Ford Motor Company adopted new machines as fast as or faster than other automakers. The new machines were often designed and built by Ford engineers, collaborating with outside machine shops, to perform only one operation and to do it continuously. Ford employees in the tool department designed the fixtures and gauges that transformed general-purpose machines into specialized ones. The special fixtures made insertion of a part into the machine automatic; the worker had to make no adjustments at all.

In 1913 production reform intensified. Individual operations were further divided, as is reflected in an account of the early flywheel magneto assembly, which was the first moving assembly. "Originally, one workman assembled the whole flywheel magneto and turned out about 40 in a 9-hour day. It was a delicate job, had to be done by an experienced man, the work was not very uniform and it was costly. In the Spring of 1913 a moving assembly line with 29 men was put into operation. The entire job was divided into 29 operations and it was found that with these 29 men, 1188 flywheel magnetos for every 9-hour day were produced, making a saving of 7 minutes time on each assembly."[27]

The magneto assembly demonstrated the principles that underlay Ford's assembly line production. "Keep the work at the least waist high, so a man doesn't have to stoop over; Make the job simple, break it up into as many small operations as possible and have each man do only one, two, or at the most three operations; Arrange so the work will come to each man so that he shall not have to take more than one step either way, either to secure his work or release it; Keep the line moving as fast as possible."[28]

Single-purpose machines and other "improved" processes further changed the allocation of work in the Ford plant. A U.S. Bureau of Labor Statistics report, published several years later, documented some of the changes: spot welding, for example, allowed one welder to do the work of eight riveters; the multiple drill press increased one man's productivity by eight to twelve times; the vertical turret lathe meant that one worker could produce sixty flywheels per day instead of the twenty-five previously machined on a regular lathe. In every case, new work organization led to increased productivity of a worker. These changes clearly required fewer workers to do the work. However, in some cases, such as the magneto assembly, workers were added as processes were divided into many small operations, and many workers were required to perform each simple operation over and over. Ford and other automakers rarely fired workers as they improved production efficiency; they simply increased overall production.[29]

The dramatic changes accomplished by 1913 had led Ford engineers to develop new ideas about how to manage production and how a factory should work. By then, specialized machines were sophisticated enough to do almost everything Ford wanted them to do, so the engineers turned their attention to production flow, focusing particularly on the handling of materials. This proved to be a more complex problem than division of labor and the use of special-purpose machines, and its importance was not overlooked: the goal of control over production and workers would not be possible without automatic flow of materials through the plant. In the following years, flow and speed of production became the most important element in the Ford Motor Company's manufacturing operation. Every facet of production, including factory design, was subordinate to production flow.

Materials Handling and Shop Floor Layout

Model T production expanded so rapidly in the Highland Park plant that it was crowded almost as soon as the company moved in. According to two of the plant's engineers, the company added new buildings as fast as it could; one building was barely finished before another was started. Such rapid growth compounded the normal problems of rational planning for the organization of production. Materials handling was especially important to such a rapidly expanding enterprise; engineers everywhere agreed that the "old-fashioned high cost of manufacturing was because of the way things were lugged around the shop."[30] In designing Highland Park, Ford and his engineers realized early on that improved internal transportation would be one of the most significant elements of fast, efficient production.

One of the novel features of the plant—the organization of the buildings—emphasized the movement of materials. Except for two processing facilities (heat treatment and foundry), the powerhouse, and the administration building, the plant's buildings shared common walls, essentially making it one large, integrated building. This contiguity meant that, as the Model T was built, parts and subassemblies traveled shorter distances, a factor that meant faster flow of materials and considerable savings in time and labor. Few factories before or after Highland Park were built in such a tight formation. Most large factories before the 1920s were built in hollow squares, or in H, L, or E configurations, shapes that allowed natural light to enter the workshops of the multiple-story buildings. Instead of leaving the usual space between buildings, the Ford Motor Company covered them with glass and transformed them into work space.

Highland Park's building arrangement eliminated costly and time-consuming travel between buildings. Except for the few factories equipped with narrow gauge railroad tracks, most companies in 1910 used horses and wagons to move materials between buildings. The expense of the horse-drawn method, which included the wages of the teamster and his helper, the costs of feed, repairs and renewals to harnesses and wagons, shoeing, and veterinary service, and the depreciation of wagons and trucks, came to approximately $1,850 a year for each team. More important than the cost were the limitations imposed by the method of transportation. In addition to the awkwardness of the large draft animals on the plant grounds, only eight to ten loads (with a two-ton limit) per team could be expected to be moved in one day.[31]

In 1910 the company's materials handling system was relatively simple. The original buildings contained only a few mechanical handling devices—two cranes, which moved materials as well as heavy equipment through the single-story machine shop, and a monorail, which moved along tracks around the entire first floor to deliver goods to areas beyond the craneways. The single-story craneway in the machine shop formed the main distributing artery in the plant. In general, parts purchased from outside, as well as materials to be shaped at the factory, were unloaded from wagons and railroad cars directly into the craneway. Next they were placed either into bins to be moved to a specific department or directly onto the craneway floor for temporary storage. Materials and parts destined for longer storage traveled by elevator to the upper

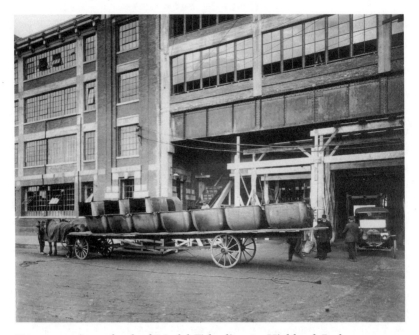

Horses moving a load of Model T bodies, to Highland Park, 1913–1914. Handling innovations arrived gradually, even at the Ford Motor Company. Neg 833.407, Ford Motor Company Archives, Dearborn, Mich.

floors.[32] From the storage areas low-paid laborers, called pushers and shovers, moved parts and materials in barrels, boxes, and pallets to the appropriate areas of the factory, removing all need for machine operators and assemblers to move from their stations in search of supplies.

As the production process changed, workers' movements around the factory were restricted because efficiency demanded that they remain at their stations. At the same time, speed and volume of production increased, and factory managers had to figure out how to get parts and components to work stations and how to get the completed subassemblies to the final assembly floor. In order to understand the significance of materials handling, visualize the bustling factory. Production rates were doubling almost every year. The 1910 factory was designed when the company produced roughly 70 cars per day, but in 1913 it was turning out between 500 and 800 per day. The work force was growing almost as fast as production: in 1910 the Highland Park had about three thousand production employees; by 1913 over fourteen thousand were on the

Monorail at Highland Park's Old Shop, demonstrating one of Ford's experiments with the interior handling of materials. This system served the first floor. Neg. 833-809, Ford Motor Company Archives, Dearborn, Mich.

payroll (in 1914 the work force was divided into three shifts instead of two); by 1917 the number of employees had increased to about thirty-six thousand.[33]

The challenge for the Ford engineers was to time and coordinate handling so that parts and materials reached a work station when needed, but not so early as to cause a backup of supplies on the shop floor. In 1912 the handling system concerned Ford and his engineers for several reasons. First, the volume of materials moving through the factory was too large to be efficiently managed; materials, parts, and subassemblies were constantly on the move. But the labor required to distribute these materials was even harder to control.

The eight hundred to one thousand "truckmen, pushers, and shovers" did not, in the eyes of Ford and his engineers, contribute directly to manufacturing. Oscar Bornholt, one of the Ford engineers, wrote in 1913 that "trucking in the machine shop is always looked upon as an unnecessary expense . . . and all its labor is nonproductive." Henry Ford con-

The drilling machine department at Highland Park's Old Shop, illustrating the clutter that typified the scene before the advent of mechanized handling. *Iron Age,* 89 (6 June 1912).

stantly sought ways to eliminate the unnecessary expense. Unproductive labor—that is, labor spent in moving materials rather than in production—continued to be a concern for the industry for decades. As late as 1934, a National Bureau of Economic Research study found that 40 percent of the auto industry's labor-saving changes came from handling materials. "One marked tendency in the modern movement for greater industrial efficiency is the effort to reduce handling through the arrangement of equipment and processes to provide straight lineflow. The serialization of processes and machines reduces inter-process handling to a minimum."[34]

Another reason for concern over handling was that pushers and shovers could not be supervised the way that stationary workers could. These mobile workers were viewed as being slow and unreliable, and managers constantly worried about the timely delivery of parts and materials. Reliable delivery was vital to efficient operation of the large factory; if workers on the shop floors ran out of parts, their idle time added up to a significant loss of money and delayed output. To help keep the system moving, a force of thirty-six clerks, called shortage chasers, "straighten[ed] out tangles of all description in the handling of materials, trac[ed] lost items which range from the smallest part to finished cars, and ferret[ed] out opportunities for improving conditions."[35] But these clerks spent their time remedying the inevitable mistakes of the poorly paid pushers and shovers.

The wages paid to the pushers and shovers accounted for a small but important portion of the company's costs. In general, labor figured as the smallest cost category in auto production, with materials, machinery, and buildings demanding the greatest outlays. The pushers and shovers were the lowest paid of all workers, earning around $1.25 per day in 1910. Nevertheless, Henry Ford was known for his efforts to reduce costs anywhere he could even if the savings amounted to only a few cents, and he was determined to eliminate as many of the unskilled, nonproductive workers as possible.[36] After examining the factory in early 1914, the industrial journalists Horace Arnold and Fay Faurote reported that "handling of materials and work in progress of finishing is now the principal problem of motor car cost reduction."[37] Not until the company moved into its next factory, the River Rouge plant, would Ford engineers begin to find lasting solutions to this problem.

Handling grew more complex every year. One Ford Motor Company policy added confusion: anything purchased had to bought as a carload (railroad car, that is)—no small purchases were allowed. "The subsequent distribution of this material through the manufacturing processes became a handling problem of unusual magnitude."[38] Then, in 1913, trying to simplify the movement and inventory of incoming parts, the company insisted that suppliers pack all small parts in boxes or cartons that could be piled on the floor. The handlers then could move the containers rather than having to fill a barrel with parts before moving them to work stations. All bins, previously repositories for all parts and materials, were eliminated. Even parts manufactured by the company were loaded into cartons or trays that held only a specified number. The standardized boxes made inventorying supplies much faster, a necessary improvement given the constantly increasing volume of production. Abell wrote in 1913 that "by reason of the change, the cost of handling stock materials is now slightly less than the expense involved in handling half the quantity of stock a year ago."[39]

Pushers and shovers still moved standard parts barrels at the Old Shop. *Factory,* Nov. 1916; photo courtesy of Hagley Museum and Library, Wilmington, Del.

Changes in the production system also exacerbated handling problems. With the move away from group assembly, the completed components had somehow to travel between work stations and then from the work stations to the final assembly area. Since production workers were confined to their individual stations, the work had to be delivered to them; thus completed components and assemblies-in-process were constantly being pushed, shoved, or carted from one work station to the next. A more logical shop floor layout and the flexibility of the more open interior spaces of the concrete buildings facilitated the human-powered movement of materials. Arnold and Faurote wrote, "It is of record that in the old Piquette Avenue days, previous to the time when any attempts at Ford shops systemization were made and chaos reigned supreme, the first systematizer found that the Ford [engine] travelled no less than 4,000 feet in course of finishing, a distance now [1913] reduced to 334."[40]

Thus, the flow of materials was directly related to the layout of the shop floor. As the plant grew more and more crowded with steadily growing numbers of machines, cars in process, and workers, plant engineers rethought layout to help with the handling problems. Shop floor layout merged with new construction as a crucial factor in the company's success, and the job of coordinating production needs with layout and construction fell to the plant layout department. The layout department determined not the types of machine to be used but how to arrange them after they had been chosen by production engineers for the most efficient operation. Layout engineers also estimated the number of pieces that would be produced by each machine and the number of machines needed. A. M. Wibel, one of the layout engineers, described the department as planning and integrating "the overall picture of production and materials handling." The layout department was concerned, in the words of one structural engineer, with the "laying out of the work," which was, according to new theories of production, the most important part of the engineers' work: "to give the greatest economy with the least expenditure of effort." M. L. Weismyer, who worked in the layout department a few years later, described his job as "cutting out little templates and laying them out on a floor plan, juggling them around to get the most efficient layout, and then making a drawing of that and selling it to the superintendent."[41]

Movements of parts and subassemblies had to be mapped out before the machinery and equipment were installed; by 1912 engineers had completed such a map, correlating all movements in production with plant layout. But things changed so fast in the plant that no layout was right for long. With each expansion, the layout department had to rearrange machinery. "Every time we would get a change of production, it would require a complete re-layout of the plant. There was a constant shifting of departments and God help you if you had a particular job like those Bullard multimatics and you ran out of room . . . before you had enough machines to do the particular job."[42]

In one of the best surviving accounts of the Ford Motor Company's rationale for changing machine layout, Oscar Bornholt, a company engineer, described the cylinder department, illustrating the way handling facilitated production speed. "Cylinders are trucked from the foundry to the border to the aisle, down which are located the machines which perform the operation." The cylinders, being light, were then carried by the operators from the end of the aisle to their machines while the machines automatically made cuts on the positioned cylinders. Cut cylinders could be moved "to each successive machine until they land in the assembly department which borders the cylinder department." The savings of the system were that: "in placing the machines according to operations it is necessary only to truck the cylinders to the first operations and after the last . . . Each operator then lays the part down in such a place and manner as to allow the next operator to pick it up and perform his operation." Under the old system, in which "cylinders were machined in departments consisting of like machines, it would be necessary to truck them to and from each department . . . It would be conservative to estimate that the cylinder would have to be trucked about twelve times."[43]

By early 1914 machines were closer together than was usual "so that there is but barely room for the workman to make his usual movement." Henry Ford himself pushed shop floor crowding. His idea, reported by the plant's construction engineer William Pioch, was to get the machines as close together as possible to save floor space.[44] The new tight arrangement left less room for materials on the shop floor. In order to open up floor space, Ford's engineers began to move the unfinished components by elevated slides, troughs, and conveyors.

The addition of these new handling devices began to change the way work in progress moved around the plant. Gravity slides made of inclined sheet iron were built next to the work stations. After completing an operation, the worker dropped the component onto the slide, "so inclined as to carry the piece by gravity to within easy reach of the next man." This saved the work of the pusher who, in the earlier scheme, moved a box of partly finished pieces by hand cart to the next station. The gravity slides also helped to speed the process. Ford engineers found that the gravity slides, chain conveyors, and the assembly line not only helped reduce labor costs but also "cleaned up" the floor, "making more room for tools and workmen where it was thought the limit of close placing of production agencies had been reached."[45]

By 1913 plant layout and handling of materials had become so important to the factory's managers and engineers that they became the primary determinants for new building design and for the company's increasing productivity. Casual layout in the early shops had worked because production was slow, the volume low, and skill rather than efficiency ruled. As production changed, the efficient flow of work through the factory was necessary to take full advantage of the new machines and minute division of labor.

For all the revolutionary innovations that took place within it, the Highland Park plant could not be regarded as a modern factory in 1913. When Ford first moved production there, he sought space large enough to hold his rapidly expanding operations. Once established in the new factory, the company's engineers set to work to make production as fast and cheap as possible. In the end they found that efficient production necessitated careful planning of the factory building. Through their efforts to streamline production, Ford's men synthesized available approaches to industrial engineering and, in doing so, developed new expectations for factory buildings.

In the end, Highland Park proved to be neither an old-style plant like Piquette nor a modern rational factory. It can be viewed in retrospect as an intermediate experiment. When Ford's engineers originally thought about building the factory, they had only the old style of production to direct them, and the old production methods were not very demanding. The systematic incorporation of innovations in production increasingly made it clear that the factory building had to be planned around production. Older factories had been planned largely according to the power source, spatial requirements, lighting, and handling; new factories had to be built with attention to the production process. Ford engineers would design the company's next building to fit the new assembly line production, and the building would prove to be an important element in the success of Fordism.

The Rational Factory: Highland Park's New Shop, 1914–1919

WHEN HISTORIANS WRITE ABOUT THE FORD MOTOR COMpany's Highland Park plant and the beginnings of mass production, they usually focus on the development of the assembly line, which began in the 1910 buildings. But the assembly line is only the prelude to the story of those important years. Just as significant is the story of the design of the 1914 buildings, built specifically to house assembly line production: much of that story is about rethinking the way all movement took place in the factory. Design and layout went hand in hand with materials handling, so as they designed the new buildings and the new handling system, Ford's engineers thought about the factory as a whole. More than any other company at the time, Ford succeeded in making the factory run as though it were a great machine.

Between 1910 and 1913 monthly production of Model Ts increased from two thousand to over fifteen thousand.[1] The auto industry had more than tripled production during those years, and the Ford Motor Company was growing faster than other companies.[2] In 1913 the company boasted an annual production of two hundred thousand automobiles, yet even that volume failed to satisfy demand. In the face of the dramatic rise in demand and production volume and the necessary increase in numbers of workers, Ford began to think about another building campaign. The decision to expand the Highland Park plant stemmed from the clear need for additional manufacturing space and also from the realization that the new buildings could be instrumental in advancing the newly introduced assembly line system. Ford and his engineers intended the new buildings to be more than a simple physical expansion. By virtue of improved design and layout, the buildings themselves would aid in the organization and control of production. Space continued to be a constant concern to the growing enterprise, but movement soon became the overwhelming priority in designing the New Shop.

Many industries preceded automobile manufacturing in their attention to materials handling and factory design. The auto industry, however, is a particularly good illustration of rationalizing innovations because so much changed so fast. The Ford Motor Company is perhaps the best example within the auto industry because it was so aggressive in using factory design and new handling technology to improve production. Also, it received more publicity than any other company. Furthermore, the Ford Motor Company archive holds abundant and accessible information about this important era of the industry's development.

Ford was not the only automaker thinking about the increasingly important role of efficient transportation of materials. For example, in 1913 at Dodge Brothers "the handling of material . . . is in accordance with a fundamental plan of rehandling parts as little as possible."[3] In the same year the Continental Motor Works' manufacturing philosophy was described. "The proper routing of material is a large saving in production, and a factory layout which will allow any department to expand without rearrangement or moving of other departments [is essential]." Articles about smaller companies indicated that these attitudes had spread widely. The Eastern Car Company arranged its shop "so that the material will travel a comparatively short distance from one department to another. The whole layout is planned with a view to reducing the handling of material to a minimum."[4] Virtually all automakers tried new methods of materials handling, but unquestionably, Ford's system achieved the greatest changes in production and received the most press.

With characteristic speed, shortly after the first experiment with moving assembly in 1913, Ford engineers met with Albert Kahn to begin work on plans for new plant buildings. By then good factory design was a standard concern for production engineers, and those at Ford made sure that their ideas were incorporated into the new design. The company's engineers, especially Edward Gray, worked more closely with Kahn than they had in designing Highland Park's first set of buildings. The early buildings soon came to be called the Old Shop as a way to distinguish them from the New Shop, those built after 1914.

In 1915, William Knudsen, head of the assembly department, reported that the New Shop had been more carefully planned than the Old Shop. The new buildings were designed with the production process and the necessary machines in mind: in other words, the engineers decided how they wanted the factory organized, planned where all of the machinery and equipment should be, and built the factory around it. In the new buildings, "all mechanical equipment was arranged for in the contract," in order to lower the cost of the building. In the early buildings, "none of this was included in the contract and continual changes, necessitating cutting, and removal of walls, in some instances ran the cost way beyond the estimates."[5]

Experience in the Old Shop helped to guide the design of the New Shop. By 1913 general dissatisfaction over the early buildings existed among shop managers. The designers of the early Highland Park buildings had worked from 1910 assumptions about production, when skilled mechanics performed their work at individual stations; at that time, when production was based primarily on static assembly, the buildings had even been innovative. The concern then was for sufficient space, good light, and convenience of moving materials, in that order, but design and layout priorities had been changing since the company moved into those buildings. The 1913 experiments hastened the change, and as Fordism developed, demands on factory design and layout grew.

In general, complaints about the Old Shop emphasized transportation, the "all important factor which influenced the length of the manufacturing cycle." Henry Ford viewed it as a key to better management of production. "If transportation were perfect and an even flow of materials could be assured, it would not be necessary to carry any stock whatsoever. The carloads of raw materials would arrive on schedule and in the planned order and amounts and go from the railway cars into production."[6]

Most of the handling in the Old Shop had been manual. That is, laborers moved parts, raw materials, and assemblies in process around the factory; assemblers and other production workers often had to carry their work from the last worker's station to their own. Such manual handling created several problems for Ford production. Most important, it was slow and unpredictable. Because it depended on human beings instead of machines, the movement was not as easy to control as the later, mechanized system. It was inefficient because it used so many men whose tasks conceivably could be eliminated. With the moving line, the well-timed and predictable delivery of materials and parts was essential. The assembly lines moved at a constant speed, and without a supply of parts, the line had to stop.

Ford's focus on quantity production complicated handling. With larger numbers of cars in production, there were more materials to move into the plant, more parts to move to work stations, more finished subassemblies to move around the shop, and more workers moving through the factory. The increasing division of labor and larger labor force resulted in more machines on the shop floor. Attention to placement of departments and their internal layout became important to managing the high volume of production traffic. By 1913, because of the company's dramatic expansion, the factory was overcrowded with machines, workers, parts, and materials. As one manager related in early 1914, "There was not floor space enough; machine tools and factory departments were not placed as the management knew they should be, and . . . truckers, pushers, and draggers engaged in needless handling of materials and works in progress."[7] The 1913 experiments with the assembly line added to the handling problems.

The introduction of the moving assembly line presaged the auto industry's future, for with it came the realization that almost all movement through the factory could be mechanized and thus closely controlled by management. The assembly line was simply one element, though the most noticed, of the larger handling system that would develop over the next few years. Mechanized materials handling provided the final piece of Fordism and created the potential for an entirely new system of production. In the new system of constantly repeated routine movements, workers' movements began to merge with those of the machine. The assembly line helped management to dictate work technique by giving the foreman the ability to control the speed of production simply by controlling the speed of the line.

The assembly line would place the greatest demands on factory organization. To Ford and his engineers, the beauty of the moving assembly line was the assurance of steady, nonstop work; the line kept moving and workers had to work along with it. Fewer unskilled laborers moved materials around the factory because mechanical handling could do much of the work faster and with greater predictability. Rationalization instilled in managers the confidence that a certain number of cars could be produced every day. The moving assembly line gave the production department the capability "to say how many cars it can build in a given time, for it knows how fast the conveyor chain moves, and how many chassis can be put on the track."[8]

The success of the moving line, however, depended on a steady supply of parts and the steady removal of finished assemblies. This was the challenge for the designers of the New Shop. The assembly line led to such a major change in production that the Ford layout department engineered a new factory to suit it better.

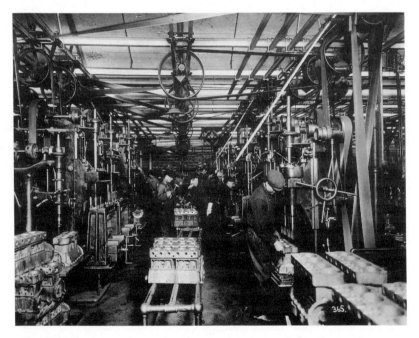

Valve reaming room with slide benches in the New Shop, Highland Park. In an early use of assembly line method, workers skid engine blocks along crude benches. Neg. 833.365, Ford Motor Company Archives, Dearborn, Mich.

The New Buildings

Although it was written up in the commercial and professional press, Highland Park's New Shop received less fanfare than the Old Shop had. It should have received more, for the New Shop took a more significant step toward factory rationalization.

Completed in 1914, the first two six-story buildings of the New Shop resembled the Old Shop in external appearance and construction style.[9] They were long and narrow (each was 62 feet wide and 842 feet long), made of reinforced concrete, and faced with brick. By 1914 the engineering of reinforced concrete had been greatly improved. Some structural weaknesses had been found in the older buildings, but the new ones would prove to be among the sturdiest of factories. Although the new buildings maintained mill building proportions, the technological and managerial innovations inside belied that traditional appearance.

Built parallel to each other, the buildings were joined by a six-story craneway that ran their entire 842-foot length. The glass-roofed craneway was the most striking, and the most important, feature of the New Shop. The craneways became the heart of the new transportation system and did much to solve handling problems. Railroad tracks, laid in the craneways, allowed the supply train to pull right into the factory. Two five-ton cranes moved along the craneway, lifting materials from the railroad cars and placing them on any one of the almost two hundred cantilevered platforms that were distributed over the six floors. Pushers and shovers then transported materials to work stations by hand truck. Because there were so many platforms, no work station was very far from the incoming raw materials. This proximity reduced the amount of movement around the factory and helped supervisors to restrict workers to their assigned stations. The system eliminated large numbers of pushers and shovers.

Cranes had been a common part of factories since the end of the nineteenth century, but the Ford Motor Company was the first to use them for the purpose of factorywide materials handling. Factories, and especially machine shops, routinely used cranes for lifting heavy materials or equipment; the Old Shop had a single-story craneway that distributed materials in its machine shop. The new cranes did much more; they changed the way the Ford engineers thought about handling. Cranes not only lifted the usual heavy materials and equipment to upper

At Ford's New Shop a moving crane, lifting materials from railroad cars below, offloaded onto platforms serving all floors. Neg. 0-1935, Ford Archives.

floors but they moved *everything* in the New Shop; all parts and materials, from car bodies to nuts and bolts, entered the New Shop via the railroad tracks and were lifted to one of the platforms by the cranes.

To make the best use of the cranes, M. L. Wiesmyer and the other layout engineers reorganized the factory and turned traditional organization on its head. In the new buildings the machine shop and foundry occupied the top floor. This change in design was a bold step on the part of the engineers. In most factories, both machine shop and foundry were typically located in a single-story building to provide good lighting (by means of saw-tooth roof) and a solid floor that could withstand the heavy machines' vibration. Placing them on the top floor was possible only because of the stronger construction of improved reinforced concrete and because of the crane's ability to lift heavy materials easily and conveniently.

The new organization allowed work in progress to descend "in natural course of operations, until it reached final assembly on the ground floor,"[10] not unlike the progression of grain through Oliver Evans's flour mill. That progression was aided by more additions to the handling system. In the New Shop, the company added chutes and ingenious spiral slides to carry subassemblies to a lower floor and improved slides to move assemblies in process from one worker to another. Eventually most parts and materials would be moved on conveyor belts.

The reorganization of production and the careful attention to flow of work in the Ford factory reflects a level of rationalization that engineers

A Model T body slides to the lower floor at the New Shop. Neg 833.28373, Ford Motor Company Archives, Dearborn, Mich.

had talked and written about for decades but few had actually achieved. To plan a factory such as the New Shop, engineers had to conceive of it as an organic whole, with every operation affecting every other. In a rational factory, engineers had to coordinate all operations and movement so that each meshed with the next. Organization and flow had been concerns in the Old Shop, but they were not planned into the design of the buildings. By 1914 engineers understood that to make the assembly line work, they had to plan for everything in the factory, that every worker and every operation required supervision. One weak link could sabotage the flow of production in the entire factory. Ford's New Shop was probably the first example of such planning in a metal fabricating industry; flow systems had worked in industries such as flour milling and meatpacking, but they were far more difficult to achieve in a plant with so many different and precise operations.

Fordism in the New Shop

In the New Shop, Ford's system of production was completed. The company could claim the highest level of rationalization realized by a manufacturer of heavy metal goods of the time. In the New Shop the three main principles of Fordism—a single, standard model, special-purpose machines, and a mechanized materials handling system—gained new importance as they were fully integrated to create a carefully thought out system of production.[11]

The Model T was crucial to Ford's initial success. The single model, produced year after year with only minor changes, opened the way for quantity production. Much of the innovation in the Ford plants has been attributed to the company's decision to produce a single model. In later years three of Ford's engineers agreed that standardization increased production. "We would not have had a shop big enough [for] our production if we had followed the old methods [producing several models]." An industrial journalist, H. F. Porter, wrote that the production of a single model allowed managers and engineers to devote their complete attention to the machines, materials, and work routines; the Model T's "unlimited run" resulted in a "free hand in selecting and developing machinery, special tools, dies, and men. Nothing is quite so demoralizing to the smooth commercial operation of a factory as incessant changes in design. Even small changes at the beginning of the season occasion much confusion for weeks or months; meanwhile, production is curtailed and

costs go skyward." Fay Faurote attributed the success of Ford's economies in production "to quantity production which, in turn, could be
attributed to the fact that only a single model was manufactured."[12]

After the decision to build only one model came experimentation
with, and introduction of, automatic specialized machines. Intense
mechanization in the shops was fundamental to the success of Fordism. Its importance stemmed from the fact that individual machines
were equipped with highly specialized fixtures and jigs that allowed the
worker to insert the part to be worked on without making adjustments.
Some machines were standard, general-purpose machines fitted with
special jigs by Ford men. Others, like the multiple screwdriving machine,
were specially designed by Ford engineers and manufactured by machine
tool companies. The screwdriving machine allowed the worker to throw
the screws "at random into the pans of hoppers at the top of the magazines." The machine placed and installed the screws, leaving the worker
nothing to do but turn the handle to move the work through the machine. While the worker operated the machine, his helper removed the
previous piece from the rack where it fell when completed.[13]

Like the standard model, specialized machines had been introduced
before the company moved into the New Shop. But because the New
Shop was planned for them, their effectiveness proved far greater than
before. The increased speed and efficiency of handling provided conditions under which the machines could work to their full capacity. "So far
as possible the piece is kept from touching the floor or from accumulat-
ing in receptacles during a series of consecutive operations in any department."[14] The system had advanced far beyond methods in the Old Shop.

Economic use of the machines depended on the scale of the factory.
According to one of Ford's engineers, most of the machines could not be
used by small manufacturers because they were available only if purchased in quantity. What is more, some were so expensive that few
companies could afford them. Another engineer, explaining that specialized and high-speed machines played an important part in the development of Ford's system, claimed that in order to build them, "you've got
to have mass production or it will break you; [they are] so expensive."[15]

In the early days of mass production, most auto manufacturers did
not have large enough output to warrant using the machines that Ford
used. However, the boom in demand for automobiles in the second
decade of the twentieth century gave other companies the opportunity to

employ some of Ford's economies of scale. By the mid-1920s descriptions of other companies sounded much like those of the Ford Motor Company. In 1926, for example, the Hudson Company made the questionable claim that it turned out more engines per hour than Ford. It had adopted Ford's practice of investing in expensive machines that "saved one minute per gear," resulting in large yearly savings. Hudson also adopted Ford's materials handling methods, moving materials quickly to avoid keeping a stockroom.[16]

The single most dramatic innovation in the Ford Motor Company's rationalization was the moving assembly line. After the introduction of the line, timing and coordination of all movement became the most important part of production management. Without improved timing and coordination, the full potential speed of the moving line would not be achieved. Consequently engineers regarded layout and handling as more important than in earlier factories. Speed became *the* production priority, and use of time a predominant concern of Henry Ford and his engineers. Workers were constantly prodded to use time well. A typical reminder in the *Ford Man*, the company paper, read:

TIME IS THE MOST VALUABLE THING IN THE WORLD

Shorten the time required to perform an operation. To save one minute in each hour worked means a saving of 1.6% in the wage bill. True efficiency means making every Minute count.[17]

Like the standard model and the special-purpose machines, the moving assembly line and other mechanized handling "saved time." That is, it forced the worker to produce more in a given amount of time. Layout and handling both addressed the constant problem—how to reduce the amount of time and labor spent in moving materials around the factory.

The changes in shop organization had the desired result. In 1915, Ford production had increased 15 percent, and the company had reduced its labor force by nearly two thousand. Two years later automobile output had been doubled in many Ford plants by rearranging the shop floor and providing an adequate supply of materials "properly timed in their arrival at the machines." In addition to the reorganized shop, a greater number of foremen increased productivity. In the new system, in which every worker had to keep up, foremen kept a close watch to ensure that parts bins were full and that everyone was working fast enough. Consequently, even though the overall number of workers declined, the ratio of foremen to workers increased from 1:53 in 1914 to 1:15 in 1917.[18]

An insurance drawing of 1919 (see fig. 5.3) illustrates the difference between building arrangement in the Old and New shops. The diagram alone cannot tell us about shop floor layout, but it is easy to see that the New Shop was organized around a linear system of handling and production. The long, narrow, side-by-side buildings of the New Shop accommodated the use of cranes as the basis of the new handling system and extended the use of the assembly line.

Vital to efficient production, layout was constrained by the line shafts that distributed mechanical power throughout the plant. Few machines ran on individual motors in the Ford factories even as late as 1918. Usually medium-sized electric engines provided power for a bank of machines, and as one engineer said, "You had to lay your departments out to utilize the line shaft overhead that drives the machine."[19]

Despite the constraints, experiments with layout did much to improve movement of materials through the factory. Continuing Old Shop practice, engineers placed machines in the New Shop as close together as possible to minimize necessary transportation between them. But that was not enough; the timing demands of the moving line also required new handling technologies. The old methods of moving parts manually created so many bottlenecks that the moving line would not be able to live up to its potential. Because it depended on human beings instead of machines, the older method was harder to control. It also required too many workers (at one time as many as 1,600) whose jobs could be eliminated. Mechanized materials handling and the new factory design were the keys to Ford's continued rationalization.

Though the cranes were the center of the New Shop's handling system, the engineers made full use of gravity slides, conveyors, and rollways. The introduction of each new technology meant less manual handling and more automatic delivery. Slides, conveyors, and rollways were built in different ways for different jobs. They varied in shape, length, width, and function. A slide might carry a finished piece just a few feet to an inspector, or it might be twenty feet long and hold the work of several men. Some slides carried work from one man to another; some moved work to the next work area, sometimes even to the next floor. Conveyor belts brought work to the worker and took it away.

Ford engineers believed that "conveyors should be installed wherever they will displace enough hand labor to pay for the investment." The in-house publication *Ford Industries* reported years later that "the day the first big conveyor went into operation seventy men were released by the transportation [i.e., handling] department for other work." Other com-panies, less convinced of the utility of conveyors, argued that "in many cases material can be conveyed in crates or boxes on trucks and elevators with great economy . . . In general, the efficiency of conveyors is apt to be overstated."[20] The critics, of course, were proven wrong as company after company adopted conveyors, and production in most factories underwent a transformation similar to that in the Ford Motor Company. The companies that continued to use the older handling system could do so because they were small-scale producers and not engaged in mass production.

Despite the obvious successes at Ford and other auto companies, not all automakers adopted the moving assembly line right away. In 1916, three years after Ford announced successful experiments with it, and after production rates in many companies had soared with the wide-spread use of the moving line, at least two prominent companies con-tinued to use group assembly. Cadillac and Chalmers continued with the old methods perhaps because they competed only on quality of product rather than on price. They too, however, eventually turned to the moving line.

It was clear that the conveyor belt would become one of the most important time-saving tools in the factory. A decade later, according to a Ford spokesman, "conveyors on which assembly or other work is done are carefully timed to insure an even output and thus act as governor on the rate of production . . . To have them move too slowly is sheer waste." But to have them move too quickly was not productive either. "Correct timing conserves the energy of the men by holding them to a uniform pace without allowing them to exceed it [and] results in a better and more uniform quality of workmanship."[21] In other words, the human machine worked best at the right pace.

The new system of conveyors also helped to solve a managerial prob-lem of increasing importance—predicting production output and its cost. Conveyors enabled the company to determine with accuracy the number of hours of labor that went into each car and every part. This permitted the factory to figure its production requirements months in advance and to regulate the flow of raw materials through the plant in such a manner that there was neither a shortage nor a surplus.[22]

Even with all the mechanical handling devices, much manual handling remained. In the New Shop, pushers and shovers, employing the standard loading boxes, continued to deliver parts and materials to the individual stations. Boxes of parts were placed along the lines to be easily accessible to the worker and to be easily seen by the foreman, who had to ensure the supply of parts.

Though we can consider it one of the first of the rational factories, Highland Park's New Shop operated under the constraints of a major holdover from earlier factory plans—the six-story buildings. Why did Ford's engineers build a multi-story building at a time when engineers and architects advocated single-story buildings for low construction costs and production efficiency? In 1914 mechanized materials handling was still in its infancy and the potential of conveyors and other devices still uncertain. It was safer, therefore, to expect that any one conveyor could reliably move materials and assemblies a few hundred feet at most. If the Highland Park plant had been spread out on one floor, it would have extended thousands of feet in each direction. Early conveyors did not have the capacity to cover such distances. O. J. Abell wrote in 1914, "It is a mistake to have single-story buildings with wide continuous floor areas, across which it is a handicap to transport material."[23] At the same time, elevators and craneways were designed for vertical movement, and they were fast, efficient, and reliable. Thus the six-story building was well suited for mechanized handling as it existed in 1914.

Work in the Rational Factory

By the end of 1914, working at "Ford's" had changed.[24] Rationalization changed almost everything in the factory. Work in Highland Park no longer resembled any other experience in a worker's life, for once the beginning bell rang, work took on an unnatural regularity. The assembly line required absolute conformity, and a worker became like one of the machines. In fact, Ford and his engineers, like many engineers before them, believed that the perfect worker would be as obedient as the machines. Years after leaving his job at Ford, one worker recalled: "Workers cease to be human beings as soon as they enter the gates of the shop. They become automatons and cease to think. They move their arms to and fro, stopping long enough to eat in order to keep the human machinery in working order for the next four hours of exploitation . . . Many healthy workers have gone to work for Ford and have come out human wrecks."[25]

The advances made in Ford's New Shop allowed the engineers to control work better. The most obvious and startling change in the entire factory was, of course, the constant movement, and the speed of that movement, not only the speed of the assembly line, but the speed of every moving person or object in the plant. When workers moved from one place to another, they were instructed to move fast. Laborers who moved parts were ordered to go faster. And everyone on a moving line worked as fast as the line dictated. Not only were workers expected to produce at a certain rate in order to earn a day's wages but they also had no choice but to work at the pace dictated by the machine. By 1914 the company employed supervisors called pushers (not the materials handlers) to "push" the men to work faster.

The 1914 jobs of most Ford workers bore little resemblance to what they had been just four years earlier, and few liked the transformation of their work. Even as early as 1912, job restructuring sought an "exceptionally specialized division of labor [to bring] the human element into condition of performing automatically with machine-like regularity and speed."[26] A good example of these changes was the piston and rod assembly. Under the old method, fourteen men worked on one bench. Each completed an entire assembly by himself—six operations—in about three minutes. The operations included: (1) drive out the pin with a special hammer; (2) oil the pin by dipping the end in a box of oil; (3) slip the pin in the rod-eye; (4) turn the pin to take the screw; (5) turn in the pinch screw; and (6) tighten the screw. Under the new regimen, the work was divided and placed on a slide bench with three men on each side. The first two drove out the pin, oiled the pin, and placed it in the piston; the second two placed the rod in the piston, passed the pin through the rod and piston, and turned the screw; the third pair tightened the screw with a wrench and placed and spread the cotter pin. The process took thirty seconds per assembly; the labor savings was 84 percent.[27]

It was no accident that as production and the factory were redesigned, work changed in a way that gave workers less control and less discretion over their work. They made fewer and fewer decisions about the operations they performed; speed and dexterity rather than knowledge of the product came to define the job. Earlier, when they assembled automobiles at their own benches, workmen controlled their own speed and could linger over a particular operation to make sure it was right. In the modern factory the shop floor layout eliminated the last vestiges of autonomy and independence in their work. One employee in Ford's Chicago plant described his work.

> Along with [the assembly line] was the other bad treatment and the fear psychoses which was developed. I worked, for example, about 40 to 50 feet from a water fountain and during the summer you would work 8:00 to noon and 12:30 to 4:30 and never be able to get over this 40 to 50 feet to get a drink of water. If you did the assembly line would move so fast you would be behind and it would be impossible to catch up. I have seen people go to the washroom and get fired when they came back because their job was behind.[28]

Workers complained about the speed and repetitiveness of the new system. The most straightforward complaint was the harsh effect of the production speed on workers' health. A nervous condition dubbed "Forditis" was attributed to the constant pressure to keep up with the pace of the line.[29] Wives also complained to the company that their husbands were overworked and physically exhausted after a day's labor.

Workers knew that the speed of the line was connected with other changes brought by rationalization, not least of which was their new role in the plant. "Workers by millions in mills and factories [were] being shaped to meet the demands of these rigid machines."[30] Jobs that once required the skills of an experienced mechanic could now be performed by any one of the new "specialists," machine tenders who knew how to perform one operation on a machine with a specialized jig. The jobs required no previous experience and a new worker could often learn the job in one day. According to a 1917 survey, by that year over 55 percent of Ford's work force were specialists. Assemblers, another semiskilled position, made up most of the balance of the productive work force. The two deskilled jobs represented the new type of worker in the auto industry.[31]

The new jobs demanded little mental activity but required instead manual dexterity, alertness, watchfulness, and rhythmic and monotonous activities, coupled with a lessening of many of the older physical requirements, rather than a knowledge of machines and the experience of earlier workers. Agility and quickness in handling parts, both large and small, became the definition of skill.[32] The shift away from skilled jobs naturally led to the decline in numbers of skilled workers in the factory. Though some of the skilled workers remained in capacities such as designers of machines and fixtures and some accepted demotion to specialist, most left the company. The new skill requirements, or lack thereof, also led to a preference for younger workers, who could perform the new operations better than old ones. In the earlier factory young men had served years as apprentices, and older men were valued for their experience. The rational factory reversed that relationship. The automatic machinery and mechanized handling equipment leveled wages by eliminating most of the highest- and lowest-paid workers.

Rationalization affected the make-up of the work force in other ways. As experienced workers left the company, untrained immigrants often replaced them. One supervisor later recalled having men of fourteen different nationalities in one department. Turnover became one of the company's most significant problems. The inventions that allowed inexperienced workers to build cars also allowed them to change jobs easily. Though turnover created disruption, the company never worried about finding enough workers to operate the machines. One historian has suggested that the jobless became "an indispensable part of rationalized industry," keeping an ever ready supply of workers outside the factory gates.[33] The factory was now filled with workers without a tradition of craftsmanship. They worked for money only.

In an account of efforts to introduce Taylorism in Watertown Arsenal in 1908, Huge Aitken described rationalized production as consisting of more than merely technological innovations. The factory represented an organization with established hierarchies, patterns of behavior, and systems of control, and introducing Taylorism necessitated widespread changes in the old patterns; "there in microcosm were all the stresses of an industrial society exposed to constant revolution in technology and organization." Aitken's observation is as pertinent to the transformation of the auto industry with the coming of Fordism as it was to the arsenal a few years earlier. However, unlike the arsenal, where workers' opposition to Taylorism resulted in a congressional ban of time study from companies holding army contracts, Ford workers had no channels for complaint and had little choice but to cooperate with the new methods or find another job.[34]

Though employees had lost their earlier position of authority over production along with the loss of their skilled jobs, they tried to assert some personal control in the rational factory in two ways. They continued soldiering practices to slow down production, but it became more and more difficult. In the new system managers could more easily identify the source of slowdowns, so they had to be planned and executed more carefully than ever before.[35] More important was the second method—unionization—which gave workers a new kind of bargaining power over working conditions and compensation.

Early efforts to organize the auto industry met with repeated difficulties. Disputes over jurisdiction characterized much of the early history; individual craft unions such as painters' and carpenters' unions competed with the Carriage Workers, who wanted to represent all workers in the industry.[36] These fights intensified in the young auto industry with the battles between the American Federation of Labor (AFL) and the Congress of Industrial Organizations (CIO) in the 1930s. But even when jurisdictional agreements were finally settled by the CIO's United Auto Workers and auto companies were organized, the union did not deal effectively with the rational factory. The rational factory contained an ironic twist in the relationship between the union and the company. On the one hand, it was easy for a small group of workers to stop the flow of production through much of the plant because of the interconnectedness of the processes. On the other hand, the very large plants of unskilled and semiskilled workers were hard to organize: the workers had no craft tradition to tie them together and little communication between departments.

Production employees were not the only ones whose jobs changed: foremen also found their jobs altered. Under rationalized production, the need grew for more careful supervision of workers and for the final checking of pieces. As work became more tedious, close supervision became necessary to assure quality control. In one way the foreman's job became easier—there were more foremen, so each one supervised fewer men. The production system also helped foremen to keep a closer eye on workers: because the special-purpose machines required standard movements, the foreman could "easily detect 'work' activities not necessary to the business at hand."[37] Likewise, foremen from other departments could monitor movement through the factory because they would know whether or not a worker had reason to be in a particular place.

The new management system resulted in a reduction in the foreman's power, however. The individual foreman in the Ford factory could no longer hire or fire workers, nor could he assign work. These functions were moved to the planning department to be performed by someone with a broad view of the entire factory. Departments were no longer the province of the foreman. Much of his earlier job—decision making about how to get production out—was usurped by engineers in the planning department. Thus the foreman lost status as he lost power to make decisions and became just another worker. His job was reduced to enforcing speed and quality of production.[38] He too became a cog in the great machine.

Highland Park's New Shop exemplifies aggressive factory planning in the second decade of the twentieth century. Drawing on nineteenth-century ideas and lessons and turn-of-the-century advances in technology, Ford engineers were able to build a rational factory whose design helped to control production much as Oliver Evans's flour mill had. No complex manufacturing process had ever achieved that kind of control before.

The New Shop's success made Ford and his engineers hungry for ever greater successes, so their experimentation continued. The constant experimentation, along with growing production volume, created an insatiable need for space. Only a year after production began in the New Shop, it was reported that "some of the buildings are now overcrowded and will, in the course of another year, need additions."[39] The desires for greater control and greater production volume combined to drive the Ford Motor Company to build its biggest plant and its greatest experiment on the banks of the Rouge River. Production continued at the Highland Park plant, and buildings were added through the 1930s to accommodate ever greater volume, but the center of the Ford operations shifted to the giant River Rouge plant.

7

Ford's Most Ambitious Machine:
The River Rouge Plant, 1919–1935

The experienced observer [sees each Rouge operation as] part
of a huge machine—he sees each unit as a carefully designed
gear which meshes with other gears and operates in synchro-
nism with them, the whole forming one huge, perfectly-timed,
smoothly operating industrial machine of almost unbelievable
efficiency.

— JOHN VAN DEVENTER, "LINKS IN A COMPLETE
INDUSTRIAL CHAIN," 1922

CONSTRUCTION AT FORD'S THOUSAND-ACRE SITE IN THE
small town of Dearborn on the banks of River Rouge began in 1917 with
the erection of huge bins for storing raw materials and traveling bridge
cranes that loaded and unloaded the bins. Work on the first blast furnace
also began in 1917. These are trifling beginnings for the plant that would
become the largest, the most expensive, and one of the most famous in
the world.

It is doubtful that in 1917 Ford and his engineers could have envisioned the industrial giant that the River Rouge plant became. By the
early 1920s, however, when the company embarked on a major building
campaign, it seems clear that they believed the plant would become the
perfect industrial machine. As they built it, they would employ all that
they and other industrial engineers had learned about efficient manufacturing, timely and predictable materials handling, and control of workers. The Rouge would take all the lessons learned at Highland Park and
expand them to the point where they were hardly recognizable.

The plant that Ford built on Detroit's Rouge River was a new kind of
factory. It was innovative not only for its size and suburban location but
also for its buildings, plant organization, and the fact that it produced

and processed almost every component of the Ford car. In the early 1920s, Henry Ford believed that his business would prosper if he no longer had to rely on outside suppliers who could raise prices for steel and other essentials as the market allowed, and he embarked on a radical venture to supply himself with raw materials. Not only did he build a steel mill, a glass plant, a rubber and tire plant, a cement plant, and others but he also bought ore and coal mines, forests, and a rubber plantation in Brazil. At the Rouge, the entire thousand-acre plant would become a great, integrated machine.[1]

When the company finally stopped building at the site decades later, even the casual observer could see that the plant was unlike other auto factories. It was so large and had so many diverse operations that it looked more like an industrial city than a factory, covering an area so great that by the 1950s it had nearly thirty miles of internal roads and over one hundred miles of railroad track for carrying parts and materials between buildings.[2]

Between 1910 and 1920, Henry Ford bought several thousand acres of land on the swampy banks of the Rouge River for himself and his company. The river connected with the Detroit River, which ran between Lake St. Clair and Lake Erie. It was not a perfect industrial site: the land would have to be filled in for building, and the river was too narrow and too shallow for the freighters that would bring materials to the plant. But it was cheap; according to Charles Sorensen, Henry Ford paid seven hundred thousand dollars for the whole site.[3]

Ford bought the land knowing that he wanted his company to keep growing and he had almost run out of room in Highland Park. Residential and commercial developments had essentially closed in the plant site. Ford and his engineers decided on the Rouge site as a solution to several problems: it would be large enough for continued expansion, water transportation through the Great Lakes would reduce high shipping costs and make delivery of raw materials more efficient; and the river would provide water necessary for proposed processing plants.

We know from some accounts that Henry Ford arranged for the purchase of some two thousand acres along the Rouge River in 1915. However, the company's corporate papers contain no mention of a purchase until November 2, 1916.[4] Two reports dated November 13, 1916 compare the desirability of the Rouge site and one on the Detroit River. The reports by William B. Mayo, one of the company's main engineers, and Julian Kennedy, the Pittsburgh engineer who would design the plant's blast furnaces, both recommended that the company choose the Rouge River location. The reports reveal some of the thoughts about the company's plans and also more general ideas about good industrial planning.

Mayo was of the opinion that "strictly from an engineering point of view," the Detroit River location would have slight advantages for a blast furnace, but taken as a whole, the River Rouge site had many more advantages. In order to construct a

> blast furnace, with connecting foundry, steel plant, and motor manufacturing plant that would be different and better than any existing plant of a similar kind and to be of such highly specialized construction to be more economical than anything yet devised[,] . . . it was necessary to pick out a location that would embrace all the necessary essentials from both the civic and the engineering viewpoint and, in addition, be so located as to have easy connection with the existing plant at Highland Park.[5]

Kennedy's report supported Mayo's. He also believed the Rouge location to be the largest convenient site near Detroit and added "It has been the universal experience of successful iron and steel works that sites which are thought to be extremely ample at the outset are found in a few years, to be entirely inadequate and more land has to be acquired . . . so that the only safe way is to start with an acreage which seems absurdly large."[6]

Mayo was enthusiastic about the Rouge site. In addition to its suitability for blast furnaces, he believed that it would allow easy access for the delivery of raw materials. He also pointed out the convenience of the location for workers and the site's proximity to "the largest labor sections in the City of Detroit." One of the most interesting of Mayo's arguments discusses the image of the new plant. Because of the "very large number of visitors," it was important to pay attention to "how well everything is done and as to its cleanliness and sanitary standards . . . that of all other things the location should be such that as near as possible a spotless town appearance could be attained both in regards to the plant and to its surroundings, with a country-like atmosphere and yet close to the city."[7]

As Mayo implied in his report, Henry Ford was also interested in Dearborn as a community. By building a factory like the Rouge in the "country-like atmosphere" of Dearborn, Ford tried to renew some of the nineteenth-century aesthetic of the factory in the country. Like those early industrialists, Ford had to reconcile a paradox: the factories he built spawned urban growth, but he personally disliked the city. He thought Dearborn would be a place where he could exercise more influence, politically and culturally, than he had in Highland Park and perhaps better control the "ills" of industrial urbanism. To that end, Ford and the company bought huge tracts of land, which eventually became sites for additional company facilities, the Edison Institute and Greenfield Village, and Dearborn Inn. The company also made plans for workers' housing in Dearborn, most of which were never carried out. Even so, Dearborn essentially became a Ford town.

World War I and the B Building

Though some work had already begun at the Rouge, World War I gave the plant its real start. Even though the Ford Motor Company was one of the fastest-growing companies in the United States, Mayo's enthusiasm for the new industrial village undoubtedly stemmed in part from the very favorable business climate in the years before the country's entry into World War I, years that were profitable and brought significant growth to U.S. industry. The decline in European manufacturing improved the commercial markets in the United States and abroad. Not only did American goods enjoy the absence of European competition but many American companies produced directly for the war effort. The prospects of the booming war economy must have had some influence on Ford's decision to buy a piece of land as large as the River Rouge site.

The new site had one very expensive problem—the river at that point was too shallow for freight ships. In December 1916, presumably as a result of the influence of the Ford Motor Company, the Army Corps of Engineers wrote the first of several reports on "improving" the Rouge River. Major H. Burgess reported that inquiries to owners of Rouge River property had been made and a public hearing held to obtain the views of all interested parties. "The result of the inquires, has been to develop the fact that the demand for further improvement of the River Rouge is due primarily to the location of a large industrial plant by the Ford Motor Company." Burgess repeatedly acknowledged that the real beneficiary of the corps's work on the river would be the Ford Motor Company, but on the basis of precedents, he argued that the improved river would provide

a public good. The public benefits would grow from increased employment (the new plant promised to employ at least 15,000) and harbor
frontage. The improved river would encourage other industries to locate
on its banks, and that, in turn, would bring greater commercial activity
(estimated at 30 million tons) to the Great Lakes. The increase of freight
traffic would be of general benefit to the city, and the turning basin,
which would allow the large ships to turn around for their return trip to
the Great Lakes (to be constructed by private concerns), would be open
to the public. Burgess clearly supported the corps's involvement with
Rouge improvements. At one point his report reads like a justification in
answer to expected criticism. "It has not been the practice of the Government to assist in the construction of private slips, but it is believed that
the conversion of the Rouge River into a slip will be of sufficient general
benefit to cause it to be considered differently from other dock improvements. It is understood that in the past it has been considered proper to
add to the harbor frontage of important cities at the expense of the
General Government."[8]

The division engineer, Colonel Frederick Abbot, opposed the improvement, arguing that it would be "in the interests of one company
and not worthy of being undertaken by the United States."[9] Nevertheless,
an agreement was reached in which the Army Corps of Engineers would
widen the channel from an irregular width (ranging from 300 to 175 feet)
to an even width of 200 feet; dredge it to a depth of 21 feet; and assume
responsibility for yearly maintenance. The estimated cost was $495,000,
plus $5,000 yearly for maintenance. In turn, the Rouge River property
owners, namely Ford, would donate land needed for widening and build
the turning basin.

The pace of construction of the early plant quickened in 1918 when the
U.S. Navy contracted with the Ford Motor Company to build Eagle
boats, which were needed after President Wilson declared the end of
U.S. neutrality in the war. Henry Ford promised to build 112 submarine
chasers within a year, and he believed that he could build them as fast as
he did Model Ts. Ford's proposal was unique in the shipbuilding world,
for no one had ever tried to mass produce ships. Ultimately the attempt
failed, but the effort had supplied Ford with a new plant.[10]

In December 1917, Henry Ford wrote to Josephus Daniels, secretary of
the navy, making an unsolicited bid to build the Eagle boats. Ford's letter
contained three promises that would later cause him trouble: first, that

his workers could easily be trained to build boats; second, that he would build them at an almost completed plant in Newark, New Jersey; and third, that he would accept no profits from the work. However, instead of using the New Jersey plant, Ford ordered the construction of a new building on the Rouge River site and billed the government $3.5 million for it. Six months after the signing of the original contract, a second was drawn up allowing the company to build a second plant at Kearny, New Jersey, at a cost of $2.5 million, to be paid by the government. In addition to the buildings, many other improvements to the Rouge, such as roads, railroad tracks, and sewage systems, were charged to the navy account. The total bill came to about $10 million.[11]

The timing of the Eagle contract proved suspiciously convenient to Ford in light of the events of the previous year. In 1916, John and Horace Dodge had filed suit against the Ford Motor Company and Henry Ford, accusing Ford of using stockholders' dividends to expand the plant and to lower the price of the Model T. They accused Henry Ford of deliberately withholding dividends and of using the money to create a monopoly on cheap cars. Though they did not explicitly argue that he wanted to hurt the Dodge Brothers Company (by then producing their own cars), that implication was clear. The suit asked that Ford distribute 75 percent of the company's cash surplus, or about thirty-nine million dollars, as dividends. In October 1917 the minority stockholders obtained a restraining order to keep the company from continuing Rouge expansion that would use the contested money.[12]

Less than two months after the court issued the restraining order, Henry Ford wrote to Daniels with his offer to build Eagle boats. Ford surely hoped that a government contract would help him to win the case and, at the very least, to nullify the restraining order. Ford and Daniels reached an informal agreement before the official contract was signed in March 1918. The government contract gave Ford the wherewithal to begin plans for the expansion of the plant. The navy would pay for the new plant, and since Ford could argue that it was necessary for the war effort, there could be no question of halting construction. Though the restraining order was lifted several months later, the navy contract allowed Ford metaphorically to thumb his nose at the Dodge brothers and the courts and proceed as planned with the new factory. Even without the Eagle contract, the company would have built the plant, but the timing gave Ford a symbolic victory plus the advantage of beginning construction while under the restraining order.

The government contract also helped the company's financial situation. In February 1919 the court ruled in favor of the Dodge brothers; the Michigan Supreme Court handed down a decision that the company owed its stockholders $19 million plus interest. Only $1.9 million went to outside owners since Henry Ford owned most of the stock. But the situation angered Ford enough that he decided to buy out all the other stockholders to ensure that they could never again keep him from reinvesting the company's profits. The buyout cost him $105 million. Not even Henry Ford had that amount of cash on hand. He borrowed $60 million, and the company liquidated as much of its inventory as possible. Without the buyout the Rouge might never have become what it was, for undoubtedly the stockholders would have continued to object to the reinvestment of profits in the plant. In the end, the United States had paid for the beginnings of the River Rouge plant and had allowed Ford to embark on his greatest industrial experiment.[13]

Construction of the plant's first building, the B building (presumably short for boat building) reached completion in 1919. One of the largest factory buildings at that time, it had three floors and measured seventeen hundred by three hundred feet, almost the size of eight football fields. Constantly improving construction technologies enabled industrial engineers and architects to design larger and larger buildings. New management techniques, along with the rapid advancements in materials handling, made efficient operations possible in such a large building. Like the Highland Park plant, this building had walls that were essentially large windows, a feature that continued in plant design through the 1920s. Unlike any of Ford's earlier factories, the building was designed to stand separate from all subsequent buildings at the Rouge plant.

The Ford Motor Company once again engaged Albert Kahn as architect for the shell of the B building. His work had developed along with the company's. By the time Kahn Associates helped build the Rouge plant, their philosophy regarding industrial architecture was fully developed: production engineers should lay out the work of the factory, and the architect "should be able to plan a factory around the scheme of production."[14]

Construction of the B building signaled Ford's total break from the

The B building, the first manufacturing structure at Ford's River Rouge plant, extended seventeen hundred feet (or nearly a third of a mile) in length. Built during World War I to make submarine chasers, it was converted to auto production as the last boat moved through assembly. Neg. 833.21863, Ford Motor Company Archives, Dearborn, Mich.

At the Rouge conveyor belts and moving overhead lines solved the problem of delivering parts and material to individual work stations. Acc. 189, Ford Motor Company Archives, Dearborn, Mich.

traditional mill building. Though in planning the New Shop at Highland Park, Ford engineers had abandoned traditional layout patterns and introduced revolutionary materials handling devices, in the end those buildings retained a similarity to the classic dimensions and appearance of the mill building. The B building retained the basic, rectangular shape of the textile mill, but its dimensions make any such comparison superficial. Its size dwarfed all factories built before it. With the old constraints of construction technology and power distribution gone, industrial engineers took advantage of new freedoms to build factories to fit production processes rather than building technologies. Engineers could lay out production any way they wanted and have a building built to fit.

With the B building, Ford engineers and architects began work on a different kind of factory. As building continued at the Rouge, individual buildings, and the plant as a whole, reflected new thinking about how to achieve rational manufacturing. The primary concern over the movement of materials continued, and the new factories would witness continued innovation in moving assemblies-in-process around the factory. Flow continued to be the key to Ford production, but the scale changed. Flow referred no longer to movement of materials only through the factory building but also around the entire thousand-acre plant *and*, we might even say, around the world as raw materials from Ford's acquisitions began to arrive at the Rouge.

Because the B building's original function was ship fabrication, the three-story interior was open, with no floors, to allow the ship to pass through the building. The original building design created problems. "The buildings were designed before the manufacturing method fully evolved, and subsequently it was found that some different set-ups might have been employed to a greater advantage. Initially the company took the view that continuous conveyor-assembly production could be applied to ship building. Further study showed that this was not practicable . . . Step by step movement was installed instead." Charles Sorensen explained in his autobiography that the building "was designed so that we could [later] add three floors the full width and use it for manufacturing and final assembly after we finished the ill-starred Eagle boats. The building was laid in to fit the ultimate plan."[15]

Postwar Production

The last Eagle boat left the B building on September 3, 1919. As it moved through the process of completion, workmen rebuilt the building's interior behind it. "As [the boats] are moved down the line from operation to operation, the installation of equipment for the building of Ford [Model T] bodies is taking place. Already the first three operations in the north end of the building are being transformed from a boat to a body building institution." The company predicted that over four thousand car bodies would be produced daily to supply to the Highland Park plant. Body manufacturing, previously done by Fisher Body and O. I. Beaudetter of Detroit, was a step toward the company's goal of integrated, self-contained production.[16] A few years later the B building also housed production of Fordson tractors.

After the war a building campaign began which did not stop until the mid-1940s, with a brief break in construction during the years of the Great Depression. Except for the B building, all the buildings during the first few years were constructed for processing raw materials—blast furnace, foundry, cement plant, powerhouse, and by-products building. Many small buildings were constructed as support for the processing plants—breakers building, screening station, pulverizing building, and others. The huge processing facilities added to the need for a plant layout different from that of Highland Park. They had different requirements, most notably for water and space. Because the processing facilities were the first to be constructed, they helped to establish the pattern for future layout.

When construction finally stopped in the 1940s, the Rouge contained almost one hundred separate buildings. These included all the buildings one would expect to find—press shop, motor building, tool and die shop, steel mill, tire plant, and so forth—as well as many that would be surprising even to the experienced industrial engineer—box factory, paper mill, waste heat power plant, benzol laboratory, and soy bean extractor building.

After the war the plant that would draw visitors from around the world began to take shape. From its conception the Rouge was dramatically different from Highland Park; as it began to grow, those differences became apparent to everyone who saw it. The Rouge is located in an industrial suburb very different from the Highland Park community. On the streets surrounding the Highland Park plant, one would have found a lively commercial area along with dense urban residential neighborhoods. The area, to use Sam Bass Warner's term, was a "walking neighborhood."[17] Ford employees walked to work, where the plant's gates opened right onto the sidewalk, offering easy access to pedestrians. Rouge workers, by contrast, could not easily walk to work because of restricted access to the plant. The situation of the Highland Park facility created a closeness between the residents and the plant which disappeared with the Rouge and later plants.

The Rouge, in fact, came to resemble a fortress; it is almost impenetrable, bounded on two sides by multiple railroad tracks, on the third by the river, and on the fourth by walls and fences. "All entrances to the main part of the Rouge are guarded when they are open and traffic is restricted to authorized vehicles and pedestrians. There is also a fence between the Fordson [tractor] yard and the parking lots and bus terminals to the east side of Miller Road and a guarded pedestrian overpass to the main part of the Rouge from the parking lots across Miller Road. There are other fences within the main part of the Rouge restricting pedestrians and vehicles to the use of guarded gates."[18]

A less striking but symbolic difference lies in a comparison of the two plants' powerhouses. Highland Park's had been built as a showplace. Not only did the stacks hold up the FORD logo, but the building's location beside a busy sidewalk invited passersby to admire the powerful and sparkling generators and beautiful fittings inside. The Rouge's powerhouse, with far greater capacity than its predecessor, was just another of the plant's buildings. The eight stacks may have symbolized the plant's size to visitors, but they did not support the company name. In fact, the Rouge could be identified by the uninitiated only by a sign at the gate that prevented visitors from freely entering the grounds. Perhaps Ford assumed that everyone would know the famous plant without clear identification. This anonymity and protective guarding is a curious contrast, however, to the company's famous tours, which allowed anyone to see the plant as part of a guided tour. The restricted access allowed the company to control entrance so that the tours gave visitors only the view

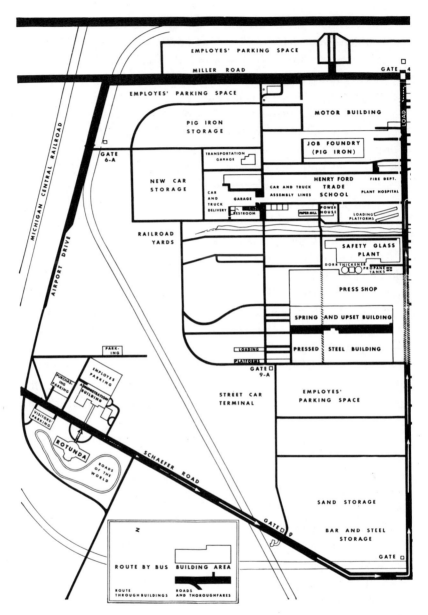

Layout of Ford's Rouge plant, covering more than 1,000 acres, at its maturity in 1941. Note the administration building and Ford Rotunda in the foreground, railroad tracks separating them from the plants themselves. Neg. 75271, Ford Motor Company Archives, Dearborn, Mich.

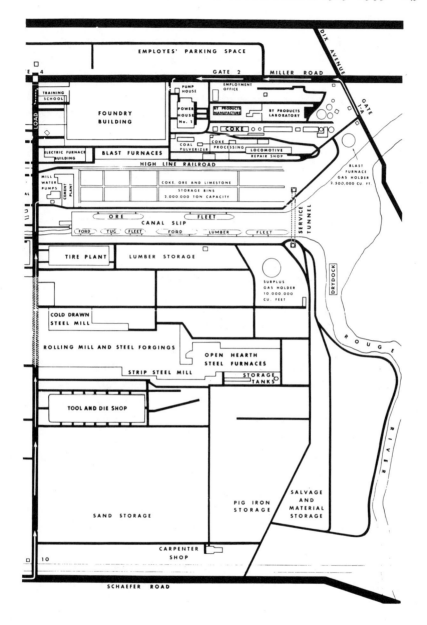

that the company approved. Ford had started his policy of plant tours at Highland Park in 1912, and the tours were so popular that the company continued them until the 1980s. Visitors from around the country and around the world viewed the operation of the famous plants, and their numbers increased with each production innovation. By the time the company stopped giving them, millions of people had seen the Ford plants.

When World War I ended, the company wasted no time moving from wartime to peacetime production. In 1919 several operations began: Model T body production got underway, the coke ovens were started, and the sawmills cut their first wood. Construction continued on the blast furnaces, which were needed in order to make the Model T's single-cast engine block. The furnaces had to be located at the Rouge because the Highland Park plant was too small and the large amount of water which would be needed was not available there.

One author has pointed out the irony of Ford's engine fabrication. Casting rather than forging the engine resulted in a product less refined though more serviceable than those of other cars. One might expect the cruder cast engine to be less expensive to manufacture when compared to the more sophisticated, high-performance engines, but, on the contrary, it required a more expensive setup. Casting required a foundry rather than a machine shop, and a stamping press and annealing furnace rather than a blacksmith and forge.[19] Hence in this case at least, the cheaper the car, the more expensive the factory. The same is true of the entire Ford plant—the most expensive factory built at the time to produce the lowest-priced car.

The first blast furnace began making iron in 1920, and it did not take long before management realized that a steel mill would be an important addition to the company's works. Charles Sorensen, who became Henry Ford's right-hand man, wrote that the company sold scrap iron for eight dollars a ton and spent more than that in handling it. The scrap iron could feasibly go right into a steel mill. But to Ford, the ability to control his own supply of steel was even more important than saving a few dollars on scrap iron. The ability to produce steel became especially important after the war, when it was a scarce commodity and auto companies competed for the available stock. As usual, Ford played a different game from everybody else. Instead of trying to outmaneuver other companies for the available steel, he simply built his own steel mill.[20]

The blast furnaces and steel mill were only the beginnings of the vast Rouge processing facilities. Over the next five years, from 1917 to 1922, the company built several more processing plants, including a glass plant, a paper mill, a cement plant, a rubber plant, a leather plant, and a textile mill. Ford took a radical step when he decided to incorporate materials processing into the production operations. But the next decision was even more unconventional—the acquisition of coal fields, forests, and rubber plantations to provide every raw material that went into the Model T. In the early 1920s the company acquired close to one million acres of forest in northern Michigan, ore mines in Michigan, coal mines in Kentucky and West Virginia, and a rubber plantation in Brazil. Wood from the forests was turned into Model T bodies. The blast furnace made ore into iron, and coal went into the steel mill. Brazilian rubber became tires. The company also experimented with making artificial rubber from soybeans for use in steering wheels.[21]

Behind the decision to invest in materials processing lay Henry Ford's frustration with and distrust of suppliers. Undependable suppliers and their costly materials had angered him in the past, and he believed that he could supply and process materials more cheaply than he could buy them. Furthermore, by controlling sources, he ensured that important materials would never be withheld because of shortages and that their prices would not be raised in tight markets.

Control over raw materials added a new dimension to the company. By building processing facilities and acquiring the sources of the materials, the Ford Motor Company entered its phase as what some have called an industrial empire. With his empire, Ford was fulfilling his dream of the totally self-sufficient plant. By adding control over all raw materials and their processing, he created the first integrated automobile factory.[22] By the time the Rouge was finished, Ford was able to watch raw materials enter the factory and a finished automobile roll out. He had succeeded in building a rational factory seemingly as efficient as Oliver Evans's flour mill.

Henry Ford's obsession with eliminating unnecessary costs led to campaigns to rid the company of all possible waste. The results showed up in additional materials processing. In order to use the by-products of one process, another was installed. I have already mentioned the scrap from the blast furnaces being turned into steel. The scrap from the sawmill and textile mill was turned into cardboard and paper for plant use. And blast furnace slag ended up as cement that the company used for all plant construction.[23]

Despite the amount of activity at the Rouge, for several years it was considered as a feeder plant to Highland Park. A 1924 company publication described the Rouge as the plant that "deals primarily in raw materials."[24] In 1927 the company changed models—from the T to the A. With the introduction of the Model A, final assembly was transferred to the Rouge. At that time the plant became the center of Ford Motor Company production.

Plant Layout and Materials Handling

The physical layout of the Rouge was less centralized than other plants. The multiple-story buildings of earlier plants, including Highland Park, had been arranged with adjoining walls or grouped around a central yard. The layout at the Rouge resembled neither of those models. To the outsider it must have looked like an almost random arrangement of huge buildings. The buildings at Highland Park had been placed contiguously to save time and expense in materials handling, a strategy that proved successful in speeding transportation at the time. The layout created another problem, though, and one not easily solved by technological innovation: expansion. With buildings so tightly arranged, each department had to be carefully planned, and if more machines or operations were added, many departments had to be shifted. This became a time-consuming and expensive operation.

When planning the Rouge, the engineers in charge of layout considered ease of expansion their major concern. By organizing the plant into distinctly separate buildings, they assured the economy and potential of future growth. An editor of *Iron Age* wrote in 1918 that at the Rouge "practically all construction is being laid out with a view of 100% expansion when necessary."[25] Engineers of the Ford Motor Company, more than their counterparts in any other auto company, had reason to believe that the almost absurd amount of space allowed for future expansion was necessary. The enormous size of the Rouge site reflects the extent and speed of the company's growth and the engineers' expectation of virtually unlimited growth. One can imagine the thrill they felt. They had the opportunity to expand the company's success at Highland Park as they developed new ideas on an unprecedented scale. Here was a company with seemingly unlimited capital resources, headed by a man willing to try almost any new idea.

Decentralization of the layout is especially apparent in considering the administration building. The first administrative offices at the Rouge were located in the plant in the Wash and Locker Building. Originally planned as a facility for workers to clean up, the building was converted into office space shortly after completion. From these offices Sorensen ran the Rouge. In 1928, the year after the introduction of the Model A, when the Rouge became the primary production facility of the company, a new administration building was built a few miles away. Though it was not far from the plant "as the crow flies," several sets of railroad tracks and a large parcel of land separating them made quick, direct travel between them impossible. Of course each manufacturing building continued to have a supervisor's office, but the company was run from the new, detached building.

The separation represented several things. Henry Ford increasingly delegated the running of the company to his son, Edsel, and to Charles Sorensen. More importantly, the company's continued expansion required an ever larger administrative and clerical staff for which physical proximity to the plant was unnecessary. Yet what was the advantage of placing the administrative offices completely away from the site?[26] The administrative offices at Highland Park were in a separate building but next to the rest of the plant. Though physically close, the offices and the plant had received different treatment. The elegance of the architectural detail set the administration building apart from the factory building, and the employees who worked there enjoyed amenities, such as the only dining room at Highland Park, which the factory workers did not have access to. And of course, office workers wore white collars rather than blue ones. The buildings reflected the differences between office work and manufacturing.

The physical separation of the Rouge's administration building from the factory reinforced the removal of decision making from the shop floor which had begun several decades earlier. The separation also reinforced the position of the industrial engineer—the engineer who looked at the factory as a whole. As manufacturing became more dependent on machines than on workers' skill, so too did production engineering become more dependent on machine design and business practice than on direct shop floor experience. The primary management of such a large operation no longer needed to be on the site. In fact, the managing engineers may have felt that the distance allowed them a symbolically better view of the whole.

Other companies preceded Ford in the separation of administration from manufacturing. Indeed, general industrial advice recommended it. The split represented the increasing professionalization of management; the new upper-level managers had less to do with the day-to-day details of manufacturing. It also reflects the growing numbers of managerial and clerical staff and the distinction between shop floor managers and business managers. By the 1920s companies such as Ford could not survive by simply making a good product. The marketing and financial decisions determined success as often as did the manufacturing operations, and many companies felt proximity to the rest of the business community to be more important than locating close to the factory. Consequently many companies moved their offices even farther from production than Ford. General Motors and Fisher Body both hired Albert Kahn to design elaborate office buildings in Detroit's new business center, a considerable distance from any manufacturing operations, and the Ford Motor Company subsequently built a skyscraper to house administrative offices even farther away from the Rouge than the original administration building.[27]

The shift to single-story buildings constituted another change in plant design which added to the new layout configuration. By 1922 single-story plants had become at least an unofficial company policy of plant construction. E. G. Liebold, Henry Ford's secretary, responded to an inquiry about plant design thus: "We find that a one-story building for factory purposes with saw-tooth construction is about the most efficient and obviates elevator service and transferring materials up and down."[28]

The Rouge was by no means the first plant to consist of single-story buildings. Many engineers had argued for years for the efficiency and economy of one-story factories because they cost less to build and provided more flexible manufacturing space. The problem with single-story plants, however, had been the difficulty in finding efficient methods by which to move materials. At Highland Park, given the state of mechanical handling knowledge, Ford's engineers found that vertical transportation was easier than horizontal. The Rouge's single-story policy worked only because technology for the movement of materials had been so much improved. Management no longer had to worry about moving parts and materials across a wide floor automatically. Horizontal transportation had become more reliable than vertical transportation.

The diffuse plant layout also worked in the efficiency-minded company because of the improved system of materials handling developed at the Rouge. Every department was equipped with mechanical handling devices, and every shop and building was connected by a network of overhead monorails and conveyors. According to Van Deventer, the development of the Rouge's "integrated manufacturing" depended largely on transportation. Integrated manufacturing "ushers in a new era of mechanical handling, announc[ing] the beginning of the exit from industry of manual lifting and shop pedestrianism, and sounds the death knell of the wheelbarrow and shovel." The continuous flow that characterized the system eliminated "any possibility of loafing or soldiering on the job when each operator is faced with the necessity of keeping up with the procession or else seeing his stock piled up to a point where it becomes distinctly noticeable by the immediate management."[29]

Process and transportation continued as the focus of changes at the Rouge. In preceding decades engineers had improved individual production machines so much that the cost of fabrication became a small part of the total cost of production. "The center of thought in the modern plant is therefore no longer the individual machine but the process . . . The biggest cost savings of today and tomorrow are likely to come from moving rather than from making. This is the decade of mechanical transportation."[30] With every innovation control over operations became tighter.

Engineers at the Rouge designed many different types of handling technologies, most of which were extensions of devices first developed at Highland Park. The moving conveyors, cranes, monorails, and railroads of earlier days persisted and were joined by overhead conveyors, the plant's unique High Line, and others. The moving conveyor continued to be a prominent feature of the company's operation. As it was improved and extended to more departments, the conveyor system continued to speed production and eliminate workers just as it had at Highland Park. An unsigned letter to Sorensen in 1929 described the consequences of new handling systems.

I cabled you today, stating that we had eliminated over 400 men in the B Building, with the installation of various conveyor systems throughout the different departments. In the torque tube department, the conveyor from the department to the shipping dock, has been completed and is in operation, eliminating 20 men for handling stock. The brake plate conveyor from this department to the loading dock [is] completed, eliminating 80 men. The differential gear case forging balcony and conveyor for same from this department to the dock, has been completed and in operation, eliminating 20 men . . . In the steel Mill, the front radius rod conveyor has been completed and in operation, eliminating 20 men. The cold heading conveyor for handling cold heading wire and finished parts will be put in operation Monday, 40 men will be eliminated.[31]

At the Rouge, engineers succeeded in almost eliminating hand trucking. Although it is listed as one of many methods still in use in 1936, it played a small role by then. The transportation around the shop floor, previously done by the hand trucks, was transferred to standard full-size, gas-powered trucks. On the upper floors of the B building there was "a truck loaded with tools, jigs, fixtures, or supplies, running along the broad aisles, stopping at certain points to unload goods and at others to pick up materials." Furthermore, the aisles are compared to streets in a "small but busy town."[32]

Overhead conveyors, made possible in the Rouge after removal of the power transmission belting, played the greatest role in ending hand trucking. By eliminating belting, pulleys, and overhead shafting, plant engineers opened the space above the machines, making the installation of the new conveyors possible. They became an important part of the Rouge production system. By maintaining a constant supply of parts for the worker, overhead conveyors did away with the necessity of storing parts at each work station; more importantly, they eliminated most trucking of parts, thereby improving reliability and speed of production.

Machine placement in the Rouge shops resembled the layout schemes that originated at Highland Park. Machines were placed closer than in more conventional shops, so close that

a man who is accustomed to the space usually allowed between machines . . . would say that the River Rouge departments were crowded and congested. Under the usual operating conditions in the average plant a machine operator is required to take many steps that have been eliminated at the River Rouge where 'the work moves and the men stand still.' An old time machinist might feel himself decidedly cramped if confined to the space allotted him in this machine shop. Inasmuch as the majority of operators at River Rouge, however, are specialists who perhaps have never even seen a machine tool before their employment by the Ford Company, they have no precedents or ingrained habits with respect to tool operation and they soon become accustomed to carrying on their operations in the space provided.[33]

The tight machine arrangement did not escape criticism, however. He may not have expressed this view while working for Ford, but years later, William Pioch, head of the tool department, criticized the practice. "It was a good idea but it didn't work out too good . . . because the machines were in so tight that sometimes if we had to move a machine, we'd have to move four or five different machines to get that one out."[34]

Engineers at the Rouge also designed a vast network of railroad tracks and roads and the "unusual High Line" to carry materials. The High Line was a concrete structure resembling a viaduct, forty feet high and wide enough to carry five railroad tracks. It has been called the "backbone of the plant" because it served as a major transportation artery throughout the facility. "The two principal functions of the High Line are active storage and distribution." The High Line was a unique mechanism for handling materials. Like the craneways of the Highland Park plant, it provided semiautomatic delivery of parts and materials to several buildings; it handled the heavy materials and transported raw materials to and from the huge storage bins. To that end, the line was equipped with hoppers and gravity unloading devices which were moved by a remote control system.[35] In keeping with its antiwaste philosophy, the company turned the area under the High Line into a one-story building housing service shops, storage, and repair stations.

In addition to the High Line, the Detroit, Toledo, and Ironton railroad, as well as the internal system of railroads and roads, moved heavy materials.[36] As at Highland Park, railroad cars could be brought right into many of the buildings to unload their cargo. "The new River Rouge plant includes the largest completely mechanized installation of handling equipment ever installed in any industrial enterprise." The handling of coal through the plant is a good example. A traveling bridge crane lifted the coal from storage bins at the ore docks and dumped it in the railroad cars on the High Line, which then carried it and dumped it on an enclosed conveyor. The conveyor carried the coal into the pulverizer building, where it was crushed to be fed into the coke ovens or oilers of the powerhouse by another conveyor.[37]

Operations at the Rouge gradually began to include manufacturing in addition to processing and partial assembly. In 1924 completion of three buildings significantly changed things. The motor plant, press building, and spring and upset building marked the beginning of the company's total move from Highland Park to the Rouge. The official explanation for the move was as follows: "Whenever production warrants it the practice of bringing the machine to the part rather than the part to the machine is followed. Eighteen Highland Park departments are being moved to the Rouge in order that all operations from the melting of ore to the final assembly of the motor may be made continuous. It is easier to move the machine to the Rouge plant than to bring castings to Highland Park in constantly increasing numbers."[38]

The company's big move to the Rouge finalized the policy of corporate centralization. The expense and technological sophistication of the plant speaks to more than centralization, however; it reflects Henry Ford's emphasis on quantity production. Though Ford himself proved a masterful publicity manager, throughout his life he encouraged technological innovation more than marketing. He believed that a good, cheap car would be its own advertisement. That thinking lies behind the construction of a complex such as Rouge.

The Ford Motor Company tradition becomes clearer when its policies are contrasted with those of General Motors. The General Motors approach concentrated energies on financial and legal skills rather than on production and mechanical ones; instead of building a company by virtue of mechanical know-how, it bought small companies along with their engineers and mechanics.

The Ford Motor Company's emergence as a leader in auto production came relatively early in the history of the automobile industry. Ford "made it" largely because he produced a cheap, good-quality car. This "first stage" of auto history, as it has been called, depended on mechanical invention and innovation. The Ford Motor Company continued to operate as though it remained in that first stage for several decades. The construction of the Rouge reflects that thinking. The organization of

General Motors, on the other hand, occurred during the "second stage" of auto history, best characterized by competition through marketing.[39]

The Model T, Ford's foremost success, stood at the center of plans for the Rouge plant. Model T sales, however, began to fall in the mid-1920s, but plant expansion continued as though the company would produce only the single model forever. The entire complex was designed around building only the Model T. Finally, in 1926, Henry Ford realized the necessity of a change of models. When design engineers finished the new Model A, it took the company six months to retool for production. This period exposed the problems with the Ford system. Single-model production meant not only that the entire plant closed down during the changeover but also that it had a built-in inflexibility that substantially slowed the retooling process. The changeover proved to be costly: the actual expense was greater than it would have been in a company organized differently, and business stopped completely during the changeover, allowing General Motors to take over the market lead. Ford continued to believe his single-model strategy was best until the market forced him to diversify. Even innovators become wedded to their own traditions.

During the changeover the company moved many operations, including final assembly, from Highland Park to the Rouge. At the same time several innovations were added. The October 1927, *Ford News,* the company paper, proudly announced that "radical advances have been made in the body department. Not a single body truck will be employed either in building a body or in transferring it to the assembly line. From first to last the body will be handled by conveyors, hoists, elevators, and transfer tables." Further improvements included some rearrangement of departments to facilitate delivery of stock. The motor assembly line was joined with the main assembly line, and assembly line practice was improved to allow the assembling of several different body types on the same line.[40]

Ford's pursuit of the rational factory had led him to build the largest, most impressive factory in the world. The River Rouge plant led the manufacturing world in innovative technology and factory design. Ford, however, had not anticipated the significant disadvantage that would accompany the successful rational factory—the lack of flexibility. The absence of adaptability hurt the company in the 1920s, 1930s, and later. The

lack of flexibility kept some manufacturers in the nineteenth and twentieth centuries from choosing the path of rationalization; they sought the
greatest adaptability possible in order to accommodate smaller, changing
markets.[41] In the end, the Rouge showed the world how far rationalization could go; it also demonstrated the disadvantages that accompanied
total rationalization.

Conclusion

In the Rouge, Ford succeeded in translating one hundred years of discourse about the rational factory into physical design. The River Rouge plant was indeed a great machine embodying the principles of efficiency, order, and control; it could produce automobiles much as Oliver Evans's mill had ground flour: the raw materials entered at one end, and a finished product left at the other. All this was done with the help of sophisticated special-purpose machines, an elaborate system of materials handling, a well-controlled work force, and a building designed for production.

The Rouge was not the unique creation of his own genius, as Henry Ford may have wanted the world to believe. Rather, it was built on the work of generations of millwrights and industrial engineers who had understood the importance of designing a building to fit the production system. They also knew that a system of moving materials through the production process efficiently and quickly was a key to economical manufacturing. Ford's engineers learned well the lessons that the earlier engineers taught through the factories they built and the books and articles they wrote. Ford engineers responded to those lessons better than most. They not only understood the message but used it aggressively as they designed expensive factories that could manufacture automobiles cheaply.

Improvements in production technology in the early twentieth century had increased the potential for faster production and larger volume. That potential could be realized, however, only with major reorganization of the factory. The Ford Motor Company, more than most companies, addressed the problems of organization by designing new factories at each stage in the development of mass production. At Ford one sees a group of engineers experiment with organizing mass production and with designing the best factory for it. Each of the company's building campaigns can be viewed as more than adding space; it should be seen as a critical point in the development of Ford's production system.

The Piquette Avenue plant represented the company's first expansion. Its builders clearly copied the nineteenth-century mill building style, which worked because the company's production process at the time differed little from nineteenth-century production. Highland Park's Old and New shops were transitional buildings, embodying characteristics of both nineteenth- and twentieth-century factories. The Old Shop represented Ford's early mass production of the Model T and the first attempts to improve the internal transportation of materials. The New Shop represented the beginning of the rational factory, a factory planned to fit the production process rather than the constraints of building technology or power transmission. The River Rouge plant undoubtedly introduced the modern rational factory to the world; the modern factory combined rational factory planning with modern production and construction technology. It demonstrated to American industry that technology and know-how had reached a level at which engineers could organize a large, totally integrated manufacturing enterprise.

The rational factory had a profound impact on the worker. It heralded the stage in industrialization at which the experience of work in the factory was totally different from nonwork life. Rationalization turned the factory into a mechanical wonder that left the worker with few authorized freedoms. Managers and engineers used factory planning and design to tackle the increasingly complicated task of organizing and controlling growing numbers of workers, machines, and materials in one place.

By the 1920s industrial engineers had made significant progress in factory planning and design. Mechanized systems of materials handling, automated machinery, and the elimination of belting for transmission of power had freed engineers to lay out the shop floor in the most efficient manner. That freedom, coupled with reinforced concrete and steel construction, opened vast possibilities for the design of factories. Mechanized handling, more than the other innovations, allowed production engineers to chart every necessary movement in the factory. With such control over details, industrial engineers succeeded in making everything in the factory merely a cog in the master machine.

In the early twentieth century, the principles behind the rational factory also appealed to professionals outside industry. Fields as far removed from manufacturing as medicine adopted engineering principles as they developed modern hospitals. Between 1910 and 1930 architects, artists, and writers, attracted to order, control, and the aesthetic appeal of the machine, created the modernist school. These nonmanufacturing spheres adopted industrial principles as they, like industrial engineers, sought order and control in an increasingly complex world. Modernist artists and writers, physicians, and other professionals all found solutions to their own challenges in the rational ideal. The factory was the clearest, most obvious use of rationalization, and it provided a vivid example of its power to order and control people and processes.

Trends in medicine and hospital organization reflected industrial values of order, control, and efficiency. Much like factory engineers, hospital administrators imagined an ideal hospital based on efficiency, discipline, and a standardized hospital routine; they talked about running their hospitals with "the perfection of a machine."[1] The hospital replaced the home or private office as the center of medical work just as the factory had replaced the workshop. Traditional medical practitioners such as midwives were replaced with trained professionals; like the engineer who replaced general supervisors, the doctor and trained technician displaced all who were not formally educated. Hospitals could be rationalized and run as efficiently as factories, patients treated as much like production goods as human beings. The latest scientific knowledge and equipment became ever more important. And as medical equipment became more complex, only the educated and specially trained personnel could use it. Medical care became more and more specialized.[2]

The architects who initially defined the modernist movement cast it in industrial terms. They looked to industrial forms as a way out of tradition-bound European architecture. The German architect Peter

Behrens recognized the significance of industrial architecture early in the twentieth century. When he built the turbine factory for Germany's AEG, the country's largest electrical manufacturer, he did not look to traditional forms but instead created a design that was boldly industrial, a design that was unromantic and clearly represented the building's function.

Other European architects followed Behrens, and they extended the industrial representations into other realms of the architectural world. Whereas Behrens had explored the essential form of industrial architecture in Europe, those who followed him developed a special fascination for U.S. agricultural and industrial structures. That fascination is best exemplified by Walter Gropius's travels through the United States and his subsequent design work. As founder of the German Bauhaus school of art and architecture as well as a respected architect. Gropius was important in defining modern architecture. His work drew on what he had found in the United States and other parts of North and South America. He was fascinated with urban factory buildings and the grain elevators that dotted the rural landscape, and he incorporated those shapes into his developing architectural style. His style rejected the decoration of earlier architecture; it portrayed the simplicity and directness of practical industrial structures.

Gropius also became an advocate of Henry Ford and F. W. Taylor, and he incorporated the principles of American mass production into architecture. By so doing, he led modern architecture beyond Behrens and rooted it in the principles of efficiency and systematization that underlay U.S. industry.[3] Their interest with Ford and Taylor led Gropius and others beyond simply using an aesthetic that drew on industrial forms; Bauhaus architects designed buildings that could actually be mass produced. They used the same principles that Ford used to mass produce cars, attending to flow, standardization of parts, and employing less-skilled workers to build mass housing.[4] Thus they sought in the building process and in their finished buildings the kind of rationalization that industrial engineers had achieved at plants such as the Rouge.

Writers also participated in the modernist movement, though their use of mechanistic themes is not as obvious as it is in art and architecture. In the work of James Joyce, however, we see clear elements of the rationalizing impulse.[5] In *Ulysses*, Joyce examined the minute details of one day, in much the same way as engineers such as the Gilbreths had examined work. He described the events of the day not as a participant but as a detached, unemotional observer undistracted by any sentimental connections to the characters.

Like the paintings of some modernist artists, Joyce's descriptions are "superrealistic." No writer had ever explored the mundane details of life as Joyce did, just as no premodernist painter had portrayed the ordinariness of machines. Joyce was dedicated to his realism. "I want to give a picture of Dublin so complete that if the city one day suddenly disappeared from the earth it could be reconstructed out of my book."[6] Consider one brief passage: "Before Nelson's pillar trams slowed, shunted, changed trolley, started for Blackrock, Kingstown and Dalkey, Clonskea, Rathgar and Terenure, Palmerston Park and upper Rathmines, Sandymount Green, Rathmines, Ringsend and Sandymount Tower, Harold's Cross. The hoarse Dublin United Tramway Company's timekeeper bawled off:— Rathgar and Terenure!"[7]

Joyce calls his realism "scrupulous meanness," that is, a conscientious attention to humble or paltry detail—not the stuff of nineteenth-century literature. "In realism you get down to the facts on which the world is based; that sudden reality which smashes romanticism into a pulp." Joyce acknowledged his work's connection with industry-influenced movements when he said to Samuel Beckett that he had "oversystematized *Ulysses*."[8] Traditional writers did not systematize; modern writers, like engineers, did.

Using mechanistic inspiration less subtly than Joyce, the American poet William Carlos Williams referred specifically to the machine. Williams strove to write "efficient" poetry, poetry with no extraneous words or parts, poetry with no redundancies. "A poem is a small (or large) machine made of words . . . there can be no part, as in any other machine, that is redundant."[9] As he developed his mature style in the second decade of the twentieth century, Williams, a physician, was influenced by the efficiency of the scientific management movement, which had moved easily into the medical community.

Artists were influenced by the machine in different ways. Perhaps this influence was best expressed by the French artist Francis Picabia: the modernists believed that "machinery is the soul of the modern world."[10] They explored the modern world through their studies of machines, factories, and a harsh, realistic treatment of nonmachine subjects.

The success of the Rouge as a rational factory is nowhere better depicted than in the modern art that portrayed the industrial giant. In 1927, as part of an advertising scheme to promote the company's new Model A, the Ford Motor Company commissioned the artist Charles Sheeler to produce a series of documentary photographs of the River Rouge plant.[11] Sheeler had already earned recognition for his urban and rural landscapes, but his industrial landscapes would become his most famous artistic achievement. Despite the original intent of the commission, the Rouge photos were hardly documentary. Instead they reflected Sheeler's ironically romantic vision of industry: a clean and orderly plant, devoid of heavy industry's smoke and grime. To achieve that mood, Sheeler chose isolated views, views that created an abstract and almost beautiful image of the plant. His photographs, like his later paintings, belied the reality of the factory; they make the viewer forget the noise, the heat, and the smells that emanated from it.

Around 1930, Sheeler began to use his photographs as the basis for a series of paintings, in which he presents the plant in a cold romanticism. "It is the industrial landscape pastoralized," removed from "the frenzied movement and clamor we associate with the industrial scene," as Leo Marx has written about Sheeler's *American Landscape*.[12] Clean, efficient buildings commanding order and respect fill Sheeler's canvases. He achieved that impression by simplifying the scenes, by stripping the factory of its complexity. He painted only major forms, and to enhance the pastoral impression he painted them in pastels. He left out the clutter of the real factory and, perhaps most revealing, he left out the workers— his fascination was for the machine. By visually removing the workers, Sheeler produced his own version of the rational factory, the factory that operated just like a machine, a factory in which workers were important only as hidden pieces of the machine, as cogs.

Sheeler's quiet landscapes provide concrete images for John Van Deventer's description of the plant. "I have said that the first impression of the River Rouge is of vastness and complexity. The second impression is that of motionless quiet. Unless one enters a building, there does not seem to be much going on. One does not see many people in the yards, perhaps the reason being that space is so vast that men are less noticeable. As a matter of fact comparatively few men are required outside the buildings because of the extent to which mechanical transportation has been developed."[13]

Sheeler, like many contemporary artists, revered "the machine." His art glorifies the factory much as earlier artists paid homage to the church. In fact, the caption beneath one photograph of the Rouge, published in *Vanity Fair* in 1928, referred to the plant in religious terms: "an American altar of the God-objective of Mass Production" and a "Mecca toward which the pious journey for prayer." A few years earlier, Van Deventer had similarly praised the plant and Henry Ford. "Steam and power from the water, glass from the sand, ore and coal from the cliffs, limestone from the gray rocks and lumber from the trees.—and thus the motor car and tractor. Henry Ford has brought the hand of God and the hand of Man closer together at River Rouge than they have ever been brought in any other industrial undertaking."[14]

Sheeler wrote that industry in the United States was replacing the church, because it affected the greatest number of people. Industry, therefore, should be the concern of the artist. "Every age manifests the nature of its content by some external form of evidence." With the decline of religious beliefs, technology had become the focus for the major belief system. Consequently, the factory took on new symbolic importance.[15]

In addition to the actual images he created, Sheeler's titles further disclose his ideas about the factory. *American Landscape*, *Classic Landscape*, *Ballet Mechanique*, and *City Interior* suggest that he believed industrial environments to be more than places for production. His titles imbue the images with a significance beyond manufacturing as he made the factory a major cultural symbol. Sheeler's landscapes, as an expression of an American attitude, say much about changes in thinking about the factory. Compared to the negative literary images of eighteenth- and nineteenth-century factories, Sheeler's visual depictions of the River Rouge plant embody an enthusiasm for industry, reflecting the impact of an industrial society on American attitudes. He captured the growing belief in industry as the messiah for modern society and in the factory as its earthly representation.

In 1931, just a few years after Sheeler began his work for Ford, Edsel Ford contracted with the Mexican artist Diego Rivera to paint murals with an industrial theme in the courtyard of the Detroit Institute of Arts. Like Sheeler, Rivera was fascinated with the River Rouge plant and chose it as the main subject for his murals. Also like Sheeler, Rivera considered modern factories important as architecture and as a cultural symbol. "In all the constructions of man's past—pyramids, Roman roads and aqueducts, cathedrals and palaces, there is nothing to equal these [factories]."[16]

The resemblance between the two artists stops with their common respect for industry and technology. Rivera's murals, in contrast to Sheeler's landscapes, step inside the buildings of the Rouge plant to the teeming world of auto production and workers and depict a voluptuous Rouge, in contrast to the cool, detached crispness of Sheeler's paintings. Rivera painted the machines with sensuous, soft, curved edges. His panels are filled with people, in contrast to Sheeler's humanless exterior images. Rivera portrayed workers and machines, though for him workers remained the most vital force in the factory.

Rivera believed in the importance of the factory in modern life, placing industrial buildings, machine design, and engineering in positions loftier than any other accomplishments in all history. But unlike Sheeler, Rivera, a Marxist, saw the worker as the center of and the power behind the great machine.[17] In focusing on people in the plant, his murals suggest a negative side of factory life. He depicts the machines as larger than life, as dominating the men around them and dictating the pace of work. The activity in the mural is almost too much for the viewer to comprehend. In stark contrast to Sheeler's serene landscape, Rivera's Rouge is frenetic, reminding one of the Forditis suffered by workers in the early days of the assembly line.

The two artists depicted different realities of the modern rational factory. Both images have their own truths. Sheeler revered the machine; to him the Rouge represented the rational factory, one with little need for workers. He perceived and praised the ideal toward which industrial engineers worked. Rivera, on the other hand, recognized some of the consequences of that ideal. He saw the ambivalent relationship of the modern factory to history, to the environment, and to workers.

It was fitting that both artists portrayed Ford's River Rouge plant, at the time the most modern of factories. The Rouge was the culmination of three decades of experimentation with factory planning and design. It seemed to represent the direction in which modern industry would move, in both building style and production methods; it was a harbinger of things to come. Above all, it reflected the growing recognition of the importance of factory organization. The Ford Motor Company, and other companies, looked to the redesign and reorganization of the factory as a way to facilitate the perfection of assembly line production and as a way to manage workers better. The new factory layout, the assembly line, and the division of labor allowed managers to maintain closer watch over workers.

Sheeler and Rivera, as part of a larger movement of artists and architects who embraced the machine as the inspiration of a new culture and a new world, saw the machine age as dynamic and full of potential to inject life into a traditional and stultified world. Modern industry and machines meant more than a means of production; they symbolized the world of the future, a world in which humans gained increasing control over the natural world and over one another. Rivera worked on the edge of that movement. He displayed a fascination with industry and machines, but his art did more than portray the machine. Sheeler, on the other hand, worked well within the movement.

These artists recognized the forces that were changing the world, a movement away from the organic, romantic nineteenth century into the machine age of the twentieth. Like the nineteenth-century shift from sun time to clock time and from artisanal work to factory work, work and much of life was ordered by the imposition of the machine. What drew so many outside the industrial world to the themes that lay behind modern manufacturing? The rational ideal appealed to people for different reasons. Some were simply fascinated by the potential perfectibility of the machine. Others were often naïvely seduced by the notion that technology could solve human problems, by the prospect of abundance (through mass production), by the magic of the machine, or by a dream of utopia where machines did the work. Still others had less humanistic motives and could not resist the promise of order and control in a world that they viewed as imperfect. These allurements continue to draw engineers and others, for good or ill, into the spell of the rational ideal.

Notes

INTRODUCTION

1. Paul M. Atkins, *Factory Management* (New York: Prentice-Hall, 1926), p. 116.
2. Billy G. Smith, *The "Lower Sort": Philadelphia's Laboring People, 1750–1800* (Ithaca, N.Y.: Cornell University Press, 1990), has demonstrated that in many urban areas there was not actually a shortage of people willing to work but that they were not trained to do industrial work.
3. Adam Smith, *The Wealth of Nations* (1776; reprint, London: Pelican Books, 1974), p. 112.
4. Steven C. Reber, "The Enlightenment Factory: Scientists and France's Royal Porcelain Manufactory, 1753–1766" (manuscript), p. 2. Also see Kenneth L. Alder, "Forging the New Order: French Mass Production and the Language of the Machine Age, 1763–1815" (Ph.D. diss., Harvard University, 1991), for discussion of the French Enlightenment and the development of the interchangeable parts manufacture.
5. Margaret Jacob, *The Cultural Meaning of the Scientific Revolution* (New York: Knopf, 1988), p. 139.
6. Andrew Ure, *Philosophy of Manufactures or An Exposition of the Scientific, Moral, and Commercial Economy of the Factory System of Great Britain*, 3d ed. (1860; reprint, New York: Burt Franklin, 1969), p. 20. Ure was by training and profession a chemist but is better known to historians for his two

books on English industry and for his series of popular scientific lectures to workingmen.

7. Ibid., pp. 15, 20–21.
8. Charles Babbage, *On the Economy of Machinery and Manufactures*, 4th ed. (1835; reprint, New York: Augustus M. Kelley, 1963), p. 121; Jacob, *Cultural Meaning*, p. 217.
9. Ibid., p. 222, quoting from William Chapman, *Address to the Subscribers to the Canal from Carlisle to Fisher's Cross* (Newcastle, 1823).
10. See Otto Mayr, *Authority, Liberty, and Automatic Machinery in Early Modern Europe* (Baltimore: Johns Hopkins University Press, 1986), for discussions of clockwork and its imagery.
11. Tench Coxe, "Address to an Assembly Convened to Establish a Society for the Encouragement of Manufactures and Useful Arts" (Philadelphia, 1787), reprinted in Michael B. Folsum and Steven D. Lubar, eds., *The Philosophy of Manufacturers: Early Debates over Industrialization in the United States* (Cambridge: MIT Press, 1982).
12. See Alfred Chandler, *Strategy and Structure* (Cambridge: MIT Press, 1962), p. 388; David Noble, *Forces of Production: A History of Production* (New York: Knopf, 1984), p. 267; Lewis Mumford, *The Myth of the Machine: The Pentagon of Power* (New York: Harcourt Brace Jovanovich, 1964), pp. 65–68. Gabriel Kolko, *The Triumph of Conservatism: A Reinterpretation of American History, 1900–1916* (New York: Free Press, 1963), p. 3.

CHAPTER ONE. RATIONALIZING PRODUCTION IN
NINETEENTH-CENTURY AMERICA

1. *Throughput* refers to the flow of materials and production through the plant, i.e., the time it takes an item in the course of production to get "through" the factory.
2. *Niles' Weekly Register* 3 (1812–13), addenda, p. 1, as quoted in Eugene Ferguson, *Oliver Evans* (Greenville, Del.: Hagley Museum, 1980), p. 13; "Oliver Evans' Philosophy," written in his own hand on the last blank page of his book *The Abortion of the Young Steam Engineer's Guide*, which he willed to his son Cadwallader Evans, reproduced in Greville Bathe and Dorothy Bathe, *Oliver Evans: A Chronicle of Early American Engineering* (1935; reprint, New York: Arno Press, 1972), p. iv.
3. Ferguson, *Evans*, pp. 11–13.
4. Thomas Cochran, *Frontiers of Change: Early Industrialism in America* (New York: Oxford University Press, 1981), p. 67; Bathe and Bathe, *Evans*, p. 12.
5. Oliver Evans, *The Young Mill-Wright and Miller's Guide* (1795; reprint, Philadelphia: Carey & Lea, 1832), pp. 201, 237.
6. Ibid., pp. 236–37.

7. Samuel Reznick, "The Rise and Early Development of Industrial Consciousness in the United States, 1760–1830," *Journal of Economic and Business History* 4 (1932): 792–93.

8. Cochran, *Frontiers of Change*, p. 67. Despite the few eighteenth-century discussions in England and Europe about the ideal factory, virtually no one had achieved the level of rationalization that Evans did in his mill.

9. John Kasson, *Civilizing the Machine* (New York: Penguin Books, 1976), p. 7.

10. Thomas Jefferson, *Notes on the State of Virginia* (1787; reprint, Chapel Hill: University of North Carolina Press, 1955), query 19, pp. 164–65; Jefferson to William Sampson, *Niles' Weekly Register*, Feb. 15, 1817, p. 1.

11. Tench Coxe, *View of the United States of America* (Philadelphia: Printed for William Hall and Wrigley & Berriman, 1794), pp. 35–40.

12. Ibid., pp. 38, 40.

13. Kasson, *Civilizing the Machine*, pp. 19–20.

14. *Niles' Weekly Register*, Apr. 12, 1817, p. 102; Carroll D. Wright, "The Factory System of the United States," *Tenth Census of the United States* (Washington, D.C., 1880), 2: 17–19.

15. Coxe, *View of the United States*, p. 55.

16. "Manufacturing Industry of the State of New York," *Hunt's Merchant's Magazine* 15 (Oct. 1846): 370–72, reprinted in Carroll W. Pursell Jr., ed., *Readings in Technology and American Life* (Oxford: Oxford University Press, 1969), p. 40.

17. *Niles' Weekly Register*, Nov. 1, 1817, p. 149; E. I. duPont, *Niles' Weekly Register*, May 10, 1817, pp. 166–67; duPont to Isaac Briggs, Dec. 30, 1815, Eleutherian Mills Historical Library, reprinted in Pursell, *Readings*, pp. 37–38.

18. See David Hounshell, *From the American System to Mass Production, 1800–1932: The Development of Manufacturing Technology in the United States* (Baltimore: Johns Hopkins University Press, 1984), p. 96.

19. Leo Marx, *The Machine in the Garden: Technology and the Pastoral Idea in America* (New York: Oxford University Press, 1964), p. 150.

20. I am grateful to Laurence Gross of the Museum of American Textile History for his help with this section.

21. *Integrated shops* refers to shops that performed all necessary operations. For example, the early textile industry divided operations between different companies—one spun the yarn, another wove the cloth, another dyed it, etc. The Waltham mills did everything in the same mill. There is no definitive source on any of the mills, but much has been written on the early textile industry, including: Thomas Bender, *Toward an Urban Vision: Ideas and Institutions in Nineteenth-Century America* (Lexington: University of Kentucky Press, 1975); Thomas Dublin, *Women at Work: The Transformation of Work and Community in Lowell, Massachusetts, 1826–1860* (New

York: Columbia University Press, 1979); Barbara Tucker, *Samuel Slater and the Origins of the American Textile Industry, 1790–1860* (Ithaca, N.Y.: Cornell University Press, 1984); Robert Weible, *The Continuing Revolution: A History of Lowell, Massachusetts* (Lowell: Lowell Historical Society, 1991).

22. Laurence Gross, *The Course of Industrial Decline: The Boott Cotton Mills of Lowell, Massachusetts, 1835–1955* (Baltimore: Johns Hopkins University Press, 1993), pp. 9–14.

23. Nathan Appleton to Samuel Appleton, Boston, Sept. 22, 1823, cited in Bender, *Urban Vision*, p. 99. See Gross, *Industrial Decline*, p. 10 for development of this idea.

24. Gross, *Industrial Decline*, p. 31.

25. Rev. Henry A. Miles, *Lowell, As It Was, and As It Is* (Lowell, Mass., 1845), p. 128, cited in Gross, *Industrial Decline*, p. 10.

26. Louis Hunter, *History of Industrial Power in the United States* (Charlottesville: University Press of Virginia, 1979), 1:433.

27. Frank Sheldon notebook, Engineering Notebooks, Division of Engineering and Industry, Smithsonian Institution.

28. For further discussion of water power see Hunter, *Industrial Power*; Anthony Wallace, *Rockdale* (New York: Knopf, 1978).

29. This section on papermaking is based on Judith McGaw, *Most Wonderful Machine: Mechanization and Social Change in Berkshire Paper Making, 1801–1885* (Princeton: Princeton University Press, 1987).

30. Ibid., pp. 195, 149, 173.

31. Ibid., pp. 224–28.

32. This section on arms making draws from Merritt Roe Smith, *Harpers Ferry Armory and the New Technology* (Ithaca, N.Y.: Cornell University Press, 1977).

33. We know now that skills changed in the new system, but they were not actually reduced until the twentieth century. See Robert B. Gordon, "Who Turned the Mechanical Ideal into Mechanical Reality?" *Technology and Culture* 29 (Oct. 1988): 744–78.

34. See M. R. Smith, *Harpers Ferry*, for an in-depth discussion of the development of the arms industry.

35. James Oakes, *The Ruling Race: A History of American Slaveholders* (New York: Knopf, 1982), p. 181; see his ch. 6 for a discussion of agricultural journals and farm management practices.

36. Ibid., pp. 156, 184.

37. Hounshell, *From the American System to Mass Production*, is a discussion of this development.

38. See esp. Alfred D. Chandler, *The Visible Hand: The Managerial Revolution in American Business* (Cambridge: Harvard University Press, Belknap Press, 1977), for a discussion.

39. See Gordon, "Mechanical Ideal into Mechanical Reality," for a discussion of skill in the arms industry.

40. Chandler, *Visible Hand*, p. 257.

41. David Brody, *The Butcher Workmen: A Study of Unionization* (Cambridge: Harvard University Press, 1964), ch. 1, develops this idea.

42. Rudolf A. Clemen, *The American Meat and Livestock Industry* (New York, 1923), pp. 122–23.

43. John R. Commons, "Labor Conditions in Slaughtering and Meat Packing," *Quarterly Journal of Economics* 19 (Nov. 1904): 5–7, 16; Frederick Law Olmsted, *A Journey through Texas,* cited in Richard G. Arms, "From Disassembly to Assembly: Cincinnati, the Birthplace of Mass Production," *Bulletin of the Historical and Philosophical Society of Ohio,* July 1959, p. 201.

44. *National Provisioner,* Nov. 21, 1903, p. 27, cited in Brody, *Butcher Workmen,* p. 5.

45. Clemen, *American Meat and Livestock Industry,* p. 121.

46. Observation made by an English traveler, Newman Hall, in Louise Carroll Wade, *Chicago's Pride: The Stockyards, Packingtown, and Environs in the Nineteenth Century* (Urbana: University of Illinois Press, 1987), p. 62; Clemen, *American Meat and Livestock Industry,* p. 126.

47. Brody, *Butcher Workmen,* p. 2.

48. Arthur Cushman, "The Packing Plant and Its Equipment," in American Meat Institute, *The Packing Industry: A Series of Lectures* (Chicago: University of Chicago Press, 1924), pp. 106–30.

49. *Swift Illustrated,* company advertising brochure, 1888, Hagley Museum and Library, Wilmington, Del.; "Mr. Armour's Paper," *National Provisioner,* Jan. 27, 1900, p. 25, cited in Brody, *Butcher Workmen,* p. 3.

50. F. W. Wilder, *The Modern Packing House* (Chicago: Nickerson & Collins, 1905).

51. Ibid., p. 25.

52. Eugene S. Ferguson, "Mechanization of the Food Container Industry in America" (paper presented at the Symposium on the History of American Food Technology, Washington, D.C., Oct. 1979), pp. 11, 13; Earl Chapin May, *The Canning Clan* (New York: Macmillan, 1937), pp. 350–51.

53. May, *Canning Clan,* pp. 29–30.

54. Mark W. Wilde, "Industrialization of Food Processing in the United States" (Ph.D. diss., University of Delaware, 1988), p. 38; E. F. Kirwan Manufacturing Company, "Catalogue of Cans, Machinery, and General House Supplies," 1890, Hagley Museum and Library Pamphlets, Wilmington, Del.

55. May, *Canning Clan.*

56. David Brody, *Steelworkers in America: The Nonunion Era* (Cambridge: Harvard University Press, 1960), p. 1.

57. William Hogan, *Economic History of the Iron and Steel Industry in the United States* (Lexington, Mass.: D. C. Heath, 1971), 1:214–15.
58. Peter Temin, *Iron and Steel in Nineteenth-Century America* (Cambridge, MIT Press, 1964), p. 167; Brody, *Steelworkers*, pp. 2–9; Hogan, *Iron and Steel Industry*, p. 215.
59. Hogan, *Iron and Steel Industry*, p. 36.
60. Jeanne McHugh, *Alexander Holley and the Makers of Steel* (Baltimore: Johns Hopkins University Press, 1980), p. 253.
61. Temin, *Iron and Steel*, p. 179.
62. Ibid., pp. 135, 164–65.
63. Mark Reutter, *Sparrows Point: Making Steel—The Rise and Ruin of American Industrial Might* (New York: Summit Books, 1988), pp. 35–38, 41.
64. Ibid., p. 39.
65. Ibid., pp. 40–41.
66. "The Present Aspect of the Labor Question," *Iron Age* 11 (May 15, 1873): 6.

CHAPTER TWO. INDUSTRIAL ENGINEERS AND THEIR "MASTER MACHINE"

1. *Iron Age* 99 (May 31, 1917): 1319; 99 (June 7, 1917): 1381.
2. *Systematic management* refers to early efforts to organize production, to take it beyond the loose managerial styles of the nineteenth century and experiment with standardized management techniques.
3. Joseph Litterer, "Systematic Management: The Search for Order and Integration," *Business History Review* 35 (1961): 470, 473.
4. Joseph Litterer, "Systematic Management: Design for Organizational Recoupling in American Manufacturing Firms," ibid. 37 (1963): 369–91.
5. There is an abundant literature on F. W. Taylor and his management system. See Hugh G. K. Aitken, *Scientific Management in Action: Taylorism at Watertown Arsenal, 1908–1915* (1960; reprint, Princeton: Princeton University Press, 1985); Samuel Haber, *Efficiency and Uplift: Scientific Management in the Progressive Movement* (Chicago: University of Chicago Press, 1964); Milton J. Nadworny, *Scientific Management and the Unions, 1900–1932* (Cambridge: Harvard University Press, 1955).
6. F. W. Taylor, "Shop Management," in *Scientific Management* (1911; reprint, New York: Harper & Bros., 1947), pp. 17–207.
7. Henry R. Towne, foreword to Taylor, *Scientific Management;* Monte Calvert, *The Mechanical Engineer in America* (Baltimore: Johns Hopkins Press, 1967), pp. 225–27.
8. George W. Light, "A Plea for the Laboring Classes," *Boston Mechanic* 4 (Dec. 1835): 239, cited in Calvert, *Mechanical Engineer*, p. 225; Thomas Egleston, "Notes of the Mechanical Engineers' Meeting," *American Machinist*, May 28, 1881, p. 8, cited in ibid., p. 226; Henry R. Towne, "The Engineer As Economist," *Transactions of the American Society of Mechanical Engineers* 7 (1886):

428, reprinted in John R. Ritchey, ed., *Classics in Industrial Engineering* (Delphi, Ind.: Prairie Publishing, 1964); Henry R. Towne "Industrial Engineering," *American Machinist,* July 20, 1905, p. 100; Coleman Sellers, "President's Address, 1886," *Transactions of the American Society of Mechanical Engineers* 8 (1886–87): 695; Calvert, *Mechanical Engineer,* p. 225.

9. F. V. Larkin, "College Training for Industrial Engineers," *Proceedings of the Society of Industrial Engineers,* 1920, p. 49, quoting Dean J. B. Johnson (source of original quotation unknown); James N. Gunn, "Cost Keeping: A Subject of Fundamental Importance," *Engineering Magazine* 4 (Jan. 20, 1901): 708.

10. "Frederick W. Taylor Dead," *Iron Age* 95 (Mar. 25, 1915): 676–78; L. P. Alford, *Henry Lawrence Gantt* (New York: Harper Bros., 1935); Paul M. Atkins, *Factory Management* (New York: Prentice-Hall, 1926).

11. MIT Catalogues, 1866–67, 1878–79, 1885–86, 1892–93, MIT Archives.

12. Ibid., 1886–90, 1894–99.

13. Howard P. Emerson and Douglas C. Naehring, *Origins of Industrial Engineering: The Early Years of a Profession* (Norcross, Ga.: Industrial Engineering and Management Press, 1988), p. 44; Daniel Nelson, "The Transformation of University Business Education," in Daniel Nelson, ed., *A Mental Revolution: Scientific Management since Taylor* (Columbus: Ohio State University Press), pp. 84–89; Emerson and Naehring, *Origins,* ch. 5.

14. Nelson, "Transformation"; Bell Crank, letter to the editor, *Wisconsin Engineer* 9 (Apr., 1905): 178, cited in Calvert, *Mechanical Engineer,* p. 232; *Technology Review* 15 (1913): 397.

15. *Technology Review* 15 (1913): 397.

16. Comments of Howard Elliot, n.d., Exhibit D, President's Papers, AC 13, folder 149, MIT Archives.

17. Letter from Charles T. Main, n.d., ibid.; Letter from Hollis French, n.d., ibid.

18. Emerson and Naehring, *Origins,* p. 1; Charles B. Going, *Principles of Industrial Engineering* (New York: McGraw-Hill, 1911), p. 1.

19. Going, *Principles,* p. 45; J. D. Louden, "Management's Use of Industrial Engineering," in H. B. Maynard, ed., *Industrial Engineering Handbook* (New York: McGraw-Hill, 1956), pp. 1–37.

20. Calvert, *Mechanical Engineer,* p. 231.

21. George M. Parks and Roger B. Collins, "200 Years of Industrial Engineering," *Industrial Engineering,* July 1976; Emerson and Naehring, *Origins;* Going, *Principles,* pp. 1–2.

22. L. W. Wallace, president, Society of Industrial Engineers, "What the Principles of Industrial Engineering Actually Accomplish When Applied by the Four Classes of Industrial Engineers," *Proceedings of the Society of Industrial Engineers,* 1920, p. 14; P. T. Sowden, "How Industrial Engineering Reduces

Production Costs," ibid., 1922, p. 125; Lee Gallway, "The Importance of Definitions to the Industrial Engineer," ibid., p. 125; Henry R. Towne, as quoted by Wallace, "Principles," p. 16.

23. Larkin, "College Training," pp. 48–49.

24. Col. Benjamin A. Franklin, "How Industrial Engineering Serves the Chief Administrator," *Proceedings of the Society of Industrial Engineers*, 1922, p. 16.

25. Edwin Layton, "Science, Business, and the American Engineer," in Robert Perruci and Joel Gerstl, eds., *The Engineers and the Social System* (New York: Wiley, 1969), pp. 51–72; Emerson and Naehring, *Origins*, p. 91.

26. A representative few are: Hugo Diemer, "The Planning of Factory Buildings and the Influence of Design on Their Productive Capacity," *Engineering News* 50 (Mar. 24, 1904): 292–94; Charles Day, *Industrial Plants: Their Arrangement and Construction* (New York: Engineering Magazine Co., 1911); Henry G. Tyrrell, *Engineering of Shops and Factories* (New York: McGraw-Hill, 1912); Frank D. Chase, *A Better Way to Build Your New Plant* (Chicago: Poole Bros., 1919); Willard Case, *The Factory Buildings* (New York: Industrial Extension Institute, 1922); P. F. Walker, *Management Engineering: The Design and Organization of Industrial Plants* (New York: McGraw-Hill, 1924).

27. Walker, *Management Engineering*, pp. 1–2.

28. H. F. L. Orcutt, "Shop Arrangement As a Factor in Efficiency," *Engineering Magazine* 20 (Oct. 1900–Mar. 1901): 719.

29. Editorial, *Engineering Record* 48 (Oct. 1903): 386.

30. Walker, *Management Engineering*, p. 2.

31. Day, *Industrial Plants*, p. 219; Henry G. Tyrrell, *Treatise on the Design and Construction of Mill Buildings and Other Industrial Plants* (Chicago: M. C. Clark, 1911), p. xv.

32. Day, *Industrial Plants*, p. 4.

33. Tyrrell, *Treatise*, p. xv.

34. Harold D. Moore, "Influence of Plant Design on Plant Efficiency," *Mechanical Engineering* 47 (Nov. 1925): 1059.

35. *Buildings and Maintenance,* Factory Management Series (Chicago: A. W. Shaw, 1915); Atkins, *Factory Management*, p. 116; Walker, *Management Engineering*, p. 68.

36. Day, *Industrial Plants*, p. 63; Going, *Principles*, p. 164; Alford, *Gantt*, p. 168.

37. Leon P. Alford, *Laws of Management As Applied to Manufacturing* (New York: Ronald Press, 1928), p. 158.

38. Tyrrell, *Engineering*, p. 42; Chase, *Better Way*, p. 3; Robert G. Valentine, "Progressive Relations between Efficiency and Consent," *Bulletin of the Society to Promote Scientific Management* 1 (Oct. 1915): 26.

39. Henry T. Noyes, "Planning for a New Manufacturing Plant," *Annals of the American Academy of Political and Social Science* 85 (Sept. 1919): 76; Diemer, "Planning of Factory Buildings," p. 293.

40. Day, *Industrial Plants*, pp. 52, 229.

41. Moritz Kahn, *The Design and Construction of Industrial Buildings* (London: Technical Journals, 1917), 11; Noyes, "Planning," p. 74.

42. Arthur G. Anderson, *Industrial Engineering and Factory Management* (New York: Ronald Press, 1928), pp. 141–42.

43. John R. Freeman to Mr. Winslow, president, United Shoe Company, Jan. 21, 1913, MC 51, box 41, correspondence file, Jan. 1913 (2), MIT Archives.

44. John R. Freeman, "Planning the New Technology," MC 51, box 41, MIT Archives.

CHAPTER THREE. THE HUMAN MACHINE: ENGINEERS AND FACTORY WELFARE WORK

1. Washington Gladden, *Tools and the Man* (Boston: Houghton, Mifflin, 1894), p. 233; H. F. L. Orcutt, "Shop Arrangement As a Factor in Efficiency," *Engineering Magazine* 20 (Oct. 1900–Mar. 1901): 718; *Iron Age* 85 (June 16, 1910): 1409.

2. A way of thinking about workers similar to the human machine, "the human motor," had been present in Europe since the second half of the nineteenth century and became popular within European industry in the twentieth century. The ideas are similar in their use of the machine metaphor for understanding and control of the human body, but *the human motor* referred more specifically to the physiology of the human body than to its productive capacity. See Anson Rabinbach, *The Human Motor* (New York: Basic Books, 1990).

3. Richard Gillespie, *Manufacturing Knowledge: A History of the Hawthorne Experiments* (Cambridge: Cambridge University Press, 1991), p. 34; Angela Nugent, "Fit for Work: The Introduction of Physical Examinations in Industry," *Bulletin of Historical Medicine* 5 (Winter 1983): 578–95.

4. Frederic S. Lee, *The Human Machine and Industrial Efficiency* (1918; reprint, Easton, Pa.: Hive Publishing, 1974), pp. v, 3. Soldiering was the deliberate decision to work less efficiently in order to relieve the pace and in some cases keep piece rates up.

5. Richard T. Dana, *The Human Machine in Industry*, (New York: Codex Book Co., 1927; reprint, Easton, Pa.: Hive Publishing, 1980).

6. Hugo Diemer, "The Planning of Factory Buildings and the Influence of Design on Their Productive Capacity," *Engineering News* 50 (Mar. 24, 1904): 294.

7. Harry Franklin Porter, "Getting Daylight into Factories," *Factory*, Sept. 1915, p. 196; O. M. Becker, "How to Increase Factory Efficiency, I: Natural Lighting," *Engineering Magazine* 51 (1916): 843.

8. E. B. Rowe and Frank B. Rae, "The Illumination of Mills and Factories with Small Units," *Electrical Review and Western Electrician*, Sept. 11, 1909, p. 503.

9. F. L. Prentiss, "Illumination of a Cleveland Factory," *Iron Age* 93 (Mar. 5, 1914): 612; C. E. Clewell, "Industrial Lighting from the New Aspect, 1," *Electrical Review and Western Electrician,* Sept. 2, 1916, p. 407; L. P. Alford and H. C. Farrell, "The Artificial Lighting of Factories," *Electrical Review and Western Electrician,* Oct. 21, 1911, p. 852; F. B. Allen, "Important Considerations in Factory Lighting," *Electrical Review and Western Electrician,* Aug. 12, 1911, p. 323.

10. George Stickney, "Electric Lighting for Industrial Plants," *Iron Age* 85 (Feb. 1910): 446; Joseph Newman, "Good Lighting from a Factory Viewpoint," *Electrical Review and Western Electrician,* Oct. 22, 1910, p. 855.

11. "Factory Lighting," *Electrical Review and Western Electrician,* Aug. 12, 1911, p. 299.

12. "Industrial Lighting," *Iron Age* 88 (Aug. 3, 1911): 281; G. G. Chapin, "The Lighting of Industrial Plants: What Developments in Electric Lighting Have Done for the Factory—Applications of the Modern Systems," ibid. 94 (Nov. 19, 1914): 1180; Alford and Farrell, "Artificial Lighting," pp. 852–53.

13. "Timely Aspects of Factory Lighting," *Electrical Review,* Sept. 1, 1917, p. 351; "Urgent Need for Improved Lighting of Industrial Plants," ibid., June 1, 1918.

14. Gillespie, *Manufacturing Knowledge,* is the most thorough treatment of the actual history of these controversial experiments.

15. John Patterson, "Altruism and Sympathy As Factors in Works Administration," *Engineering Magazine* 20 (Oct. 1900–Mar. 1901): 579; Charles L. Hubbard, "What It Pays to Know about Heating and Ventilating," *Factory,* July 1918, p. 24.

16. Library of Factory Management, *Buildings and Upkeep* (Chicago: A. W. Shaw, 1915), pp. 135–36; Hubbard, "Heating and Ventilating," p. 25.

17. Library of Factory Management, *Buildings and Upkeep,* p. 137.

18. Gail Cooper, "Manufacturing Weather: A History of Air Conditioning in the United States" (Ph.D. diss., University of California, Santa Barbara, 1987); "Heating and Ventilating a Large Factory," *Iron Age* 89 (April 4, 1912): 844–45; Library of Factory Management, *Buildings and Upkeep,* ch. 13.

19. See Cooper, "Manufacturing Weather"; also Hubbard, "Heating and Ventilating," p. 24.

20. The technique is described in Frank Gilbreth and Lillian Gilbreth, *Applied Motion Study* (New York: Sturgis & Walton, 1917).

21. Ibid., p. 52.

22. Many scholars have written about welfare capitalism. A few useful works are: Stuart D. Brandes, *American Welfare Capitalism, 1880–1940* (Chicago: University of Chicago Press, 1976); Sanford Jacoby, *Employing Bureaucracy: Managers, Unions, and the Transformation of Work, 1900–1945* (New York: Columbia University Press, 1985); David Brody, "The Rise and Decline of

Welfare Capitalism," in *Workers in Industrial America: Essays on the 20th-Century Struggle* (New York: Oxford University Press, 1980).

23. Magnus W. Alexander, "Important Phases of the Labor Problem," *Iron Age* 102 (Nov. 1918): 1258–61, 1322–25; Magnus Alexander, "The Labor Problem Analyzed" (address to joint session of the National and American Cotton Manufacturers' associations, New York, May 1918); J. A. Estey, *The Labor Problem* (New York: McGraw-Hill, 1928).

24. Orcutt, "Shop Arrangement," pp. 717–20.

25. Charles B. Going, "Village Communities of the Factory, Machine Works, and Mine," *Engineering Magazine* 21 (1901): 59, 61.

26. Stanley Buder, *Pullman: An Experiment in Industrial Order and Community Planning, 1880–1930* (Oxford: Oxford University Press, 1967), chs. 5, 12.

27. N. O. Nelson, "The Human Factor," *Factory*, Feb. 1915, pp. 101–2.

28. John R. Schleppi, " 'It Pays': John H. Patterson and Industrial Recreation at the National Cash Register Company," *Journal of Sport History* 6 (Winter 1979): 25; Patterson, "Altruism and Sympathy," p. 577; U.S. Department of Labor, Bureau of Labor Statistics, *Employers' Welfare Work* (Washington, D.C.: Government Printing Office, 1913); "United States Steel Corporation Welfare Expenditures, Jan. 1, 1912–Dec. 31, 1925)," Bulletin no. 11, United States Steel Corporation (Dec. 1925).

29. "Views on Labor Problems Vary Widely," *Iron Age* 103 (Mar. 1919): 619; "Study Professional Side of Labor Problem," and "Home-Owning Best Antidote for Radicalism," both in ibid. 106 (Nov. 1920): 1338; Robert Hadfield, "Why Fight Labor? A Better Basis for Industrial Peace," *Factory*, June 1919, p. 1157.

30. "Free Hot Soup in Winter Combats the Saloon Free Lunch," *Factory*, Sept. 1916, p. 261; "What is Done with the Noon Hour," ibid., Sept. 1915, p. 254.

31. William Tolman, *Social Engineering* (New York: McGraw Publishing, 1909), pp. 72–90; George Price, *The Modern Factory* (New York: Wiley, 1914), p. 320.

32. Price, *Modern Factory*, pp. 315–16.

33. Robert Goldman and John Wilson, "The Rationalization of Leisure," *Politics and Society* 7 (1977): 157–87.

34. Tolman, *Social Engineering*; Patterson, "Altruism and Sympathy," pp. 577–602; U.S. Department of Labor, *Employers' Welfare Work*; *Industrial Relations* (Rochester, N.Y.: Eastman Kodak, [c. 1919]).

35. Jacoby, *Employing Bureaucracy*, p. 101.

36. William Tolman, "The Social Engineer, a New Factor in Industrial Engineering," *Cassier's Magazine*, June 1901, pp. 91–107.

37. Stephen Scheinberg, "Progressivism in Industry: The Welfare Movement in the American Factory," *Canadian Historical Association Annual Report*, 1967, p. 186–88; also see Tolman, "Social Engineer."

38. Tolman, *Social Engineering*.
39. The position of social engineer was often filled by a woman. Because social engineers dealt solely with the "human" side of workers, women were deemed suitable to do the work. They were also cheaper.
40. Tolman, *Social Engineering*, p. 2.
41. See Jacoby, *Employing Bureaucracy*, ch. 4, for discussion of thinking about stabilizing the work force and reducing unemployment.
42. Ibid., p. 101.
43. Gladden, *Tools*, p. 213.
44. See for example Henry G. Tyrrell, *Treatise on the Design and Construction of Mill Buildings and Other Industrial Plants* (Chicago: M. C. Clark, 1911); Charles Day, *Industrial Plants: Their Arrangement and Construction* (New York: Engineering Magazine Co., 1911); *Buildings and Maintenance*, Factory Management Series (Chicago: A. W. Shaw, 1915).
45. Library of Factory Management, *Buildings and Upkeep*, p. 135.
46. See discussion following the address by Col. Benjamin A. Franklin, "How Industrial Engineering Serves the Chief Administrator," *Proceedings of the Society of Industrial Engineers*, 1922, pp. 33–34.
47. Jacoby says that Taylor and some supporters opposed welfare work; see Jacoby, *Employing Bureaucracy*, pp. 54, 61.
48. Diemer, *Planning of Factory Buildings*, p. 332.
49. See Abraham Epstein, "Industrial Welfare Movement Sapping American Trade Unions," *Current History* 25 (July 1926): 516–22; Price, *Modern Factory*; George C. Nimmons, "Modern Industrial Plants," pts. 1 and 2, *Architectural Record* 45 (1918): 414–21, 532–49.
50. Epstein, "Industrial Welfare Movement," p. 522.
51. *Proceedings of the Fourth Biennial Convention of the Amalgamated Clothing Workers of America*, Cincinnati, May 14–19, 1928, p. 102, cited in Goldman and Wilson, "Rationalization of Leisure," p. 179.
52. Cited in Price, *Modern Factory*, p. 294.

CHAPTER FOUR. MODERNIZING FACTORIES IN THE EARLY
TWENTIETH CENTURY

1. "The Arrangement and Construction of a Modern Manufacturing Plant," *Iron Age* 81 (Apr. 9, 1908): 1160 (emphasis in original).
2. Horace A. Arnold, "Machine-Shop Economics," *Engineering Magazine* 11 (Apr.–Sept. 1896): 267.
3. Frederick A. Waldron, "Mechanical Transportation in the Modern Machine Shop," *Engineering Magazine* 28 (Oct. 1904–Mar. 1905): 489–90.
4. Robert Thurston Kent, "Possible Economies in Shop Transportation," *Iron Age* 92 (Aug. 7, 1913): 280–82; Harry C. Spillman, "Methods of Handling Materials in Shops," ibid. (Dec. 4, 1913): 1272.

5. Kent, "Possible Economies," pp. 280–82; "Handling Cost Reduced by Tractors," *Iron Age* 92 (July 3, 1913): 14.

6. Kent, "Possible Economies," p. 282.

7. Henry M. Wood, "Shop System for Greater Output," *Iron Age* 88 (Aug. 3, 1911): 268, 269.

8. "Handling Cost Reduced," pp. 14–15.

9. "A Conveyor System Utilizing Gravity," *Iron Age* 92 (Sept. 4, 1913): 499–500; "Labor-Saving Conveyors Used in a Factory," ibid. (Aug. 7, 1913): 275–77; "Cutting Factory Costs with Conveyor System: Gravity and Power Equipment Installed by the National Acme Manufacturing Company," ibid. 91 (Feb. 6, 1913): 349.

10. Arnold, "Machine-Shop Economics," p. 266.

11. "A Whiting Electric Railroad Repair Shop Crane," *Iron Age* 83 (Apr. 8, 1909): 1129.

12. Robert L. Streeter, "Handling Materials in Manufacturing Plants," *Engineering Magazine* 50 (1915–16): 226–33.

13. R. M. Kinny, "Saving $200 a Day in Inter-Shop Haulage," *Factory*, May 1919, pp. 926–27; Arnold, "Machine-Shop Economics," p. 266.

14. For a discussion of the history of reinforced concrete, see Carl W. Condit, *American Building* (Chicago: University of Chicago Press, 1968); Reyner Banham, *A Concrete Atlantis: U.S. Industrial Building and European Modern Architecture, 1900–1925* (Cambridge: MIT Press, 1986).

15. Condit, *American Building*, p. 168.

16. David Hounshell, *From the American System to Mass Production, 1800–1932: The Development of Manufacturing Technology in the United States* (Baltimore: Johns Hopkins University Press, 1984), ch. 6.

17. See Betsy Bahr, "New England Mill Engineering: Rationalization and Reform in Textile Mill Design, 1790–1920" (Ph.D. diss., University of Delaware, 1987), for a thorough discussion of the engineering of slow-burn buildings.

18. Robert A. Cummings, "Proposed Methods for Reinforcement of Concrete Compression Members," box 27, Cummings Papers, Division of Engineering and Industry, Smithsonian Institution.

19. Robert Cummings, speech to Engineers Society, May 17, 1910, ibid.; For more in-depth discussion of concrete structures, see David Billington, ed., *Perspectives on the History of Reinforced Concrete, 1904–1941* (Princeton: Princeton University, Department of Civil Engineering, 1980).

20. Albert Kahn, "Architectural Pioneers in Development of Industrial Buildings," *The Anchora of Delta Gamma*, n.d., p. 377, Albert Kahn Papers, Detroit Institute of Arts.

21. This describes the simplest arrangement; variations can be found in larger factories; see Louis C. Hunter, *A History of Industrial Power in the United States, 1780–1930* (Charlottesville: University Press of Virginia, 1979).

22. Charles T. Main, "Choice of Power for Textile Mills," *Proceedings of the National Association of Cotton Manufacturers,* Apr. 28, 1910, p. 1; Warren D. Devine Jr., "From Shafts to Wires: Historical Perspective on Electrification," *Journal of Economic History,* June 1983, pp. 360, 363; *American Machinist,* Feb. 14, 1901, p. 176; R. E. B. Crompton, "Electrically Operated Factory," *Cassier's Magazine,* Jan. 1896, p. 292; Stephen Green, "Modifications in Mill Design Resulting from Changes in Motive Power," *Transactions of New England Cotton Manufacturers Association* 63 (Oct. 1897): 128–36.

23. R. T. E. Lozier, "Direct Electrical Drive in Machine Shops," *Cassier's Magazine,* Jan. 1896, p. 291–89; Devine, "From Shafts to Wires," p. 366.

24. Devine, "From Shafts to Wires."

25. Michigan, Bureau of Labor Statistics, 21st Annual Report (Lansing, 1904), p. 208.

26. Ibid., pp. 209, 213; George S. May, *R. E. Olds* (Grand Rapids, Mich.: Eerdmans Publishing, 1977), p. 127.

27. Grant Hildebrand, *Designing for Industry: The Architecture of Albert Kahn* (Cambridge: MIT Press, 1974), p. 28; W. Hawkins Ferry, *The Buildings of Detroit* (Detroit: Wayne State University Press, 1980).

28. Michigan, *Bureau of Labor Statistics, 21st Annual Report,* p. 208; Fred Colvin to Sidney Miller, Nov. 12, 1926, Acc. 96, Dodge Estate, box 3, Ford Motor Company Archives, Edison Institute, Dearborn, Mich. (henceforth referred to as Ford Archives).

29. For details on the way this assembly group worked, see Hounshell, *From the American System to Mass Production,* ch. 6. The transfer of the arms-type manufacturing system from New England to the Detroit auto industry is the subject of the latter chapters of Hounshell.

30. Joyce Shaw Peterson, *American Automobile Workers* (Albany: State University of New York Press, 1987), p. 46.

31. John W. Anderson to his father, June 4, 1903, Acc. 1, box 114, Ford Archives; Allan Nevins, *Ford: The Times, the Man, the Company,* vol. 1 of Nevins, *Ford* (New York: Scribner's, 1954), p. 228.

32. The lot was 430 feet by 380 feet and the building filled much of it. The architects were Field, Hinchman, and Smith. See Banham, *Concrete Atlantis,* for an excellent discussion of the daylight factory.

33. Nevins, *Ford: The Times,* p. 364.

34. It was established as a separate company as a tactic to buy out a major stockholder (ibid., pp. 278–79).

35. See Hounshell, *From the American System to Mass Production,* ch. 2, for discussion.

36. Conference with P. Martin, Hartner, and Dagner, June 3, 1926, re Additional Tax Case, Acc. 96, box 12, Ford Archives; Hounshell, *From the American System to Mass Production,* pp. 220–22; May, *Olds,* p. 192.

37. Good evidence regarding the additional buildings and the move is not available. Nevins, *Ford: The Times*, ch. 14, talks about Piquette expansion, and photographs in the Ford Archives show the buildings in the expansion. The photographs suggest strongly that the additional buildings were not built for the plant but already existed on the site.

CHAPTER FIVE. THE CRYSTAL PALACE: THE FORD MOTOR COMPANY'S
HIGHLAND PARK PLANT, 1910–1914

1. Descriptions of these and other auto plants are found in *Iron Age*, vols. 85–86 (1910).

2. Alfred D. Chandler, *The Visible Hand* (Cambridge: Harvard University Press, Belknap Press, 1977), p. 243.

3. Reyner Banham, *A Concrete Atlantis: U.S. Industrial Building and European Modern Architecture, 1900–1925* (Cambridge: MIT Press, 1986).

4. Grant Hildebrand, *Designing for Industry: The Architecture of Albert Kahn* (Cambridge: MIT Press, 1974), p. 44; Stephen Meyer, *The Five Dollar Day: Labor Management and Social Control in the Ford Motor Company, 1908–1921* (Albany: State University of New York Press, 1981), p. 24.

5. William Tolman, *Social Engineering* (New York: McGraw Publishing, 1909), ch. 3; George Price, *The Modern Factory* (New York: Wiley, 1914), ch. 5; O. M. Becker, "How to Increase Factory Efficiency, I—Natural Lighting," *Engineering Magazine* 51 (1916): 835–52.

6. Banham, *Concrete Atlantis*.

7. Correspondence between Robert Cummings and Taylor-Wilson Manufacturing Co., June 21, 26, 1905, box 14, Cummings Papers, Division of Engineering and Industry, Smithsonian Institution.

8. "B. R. Brown Reminiscences" and R. T. Walker Reminiscences," Ford Archives.

9. Allgemeine Elektricitäts-Gesellshaft was Germany's equivalent to General Electric.

10. Tilmann Buiddensieg, *Industriekultur: Peter Behrens and the AEG, 1907–1914* (Cambridge: MIT Press, 1984), p. 2.

11. Albert Kahn, "Architect Pioneers in Development of Industrial Buildings," *The Anchora of Delta Gamma*, n.d., pp. 376–78, Albert Kahn Papers, Detroit Institute of Arts.

12. "George Thompson Reminiscences," Ford Archives.

13. Dodge Case, 1917, Acc. 33, box 41, Ford Archives.

14. David L. Lewis, "Ford and Kahn," *Michigan History* 64 (Sept.–Oct, 1980): 17.

15. E. R. Breech Papers, Acc. AR 65-71:22, Ford Industrial Archives, Redding, Mich.

16. Hildebrand, *Designing for Industry*, p. 44.

17. Insurance Appraisal, vol. 1, box 8, Acc. 73, Ford Archives.

18. Ibid.

19. Memo from Detroit Architectural Iron Works, Acc. 361, box 7, Ford Archives; also "R. T. Walker Reminiscences," p. 10, and "William Vernor Reminiscences," p. 3, Ford Archives.

20. "New Works for Manufacture of Engineering Specialties," *Electrical Review,* Nov. 15, 1902, pp. 689–90.

21. Philip S. Foner, *History of the Labor Movement in the United States* (New York: International Publishers, 1965), 4:385.

22. See Allan Nevins, *Ford: The Times, the Man, the Company,* vol. 1 of Nevins, *Ford* (New York: Scribner's, 1954); David Hounshell, *From the American System to Mass Production, 1800–1932: The Development of Manufacturing Technology in the United States* (Baltimore: Johns Hopkins University Press, 1984); Meyer, *Five Dollar Day.*

23. Hounshell, *From the American System to Mass Production,* pp. 235–37.

24. Horace L. Arnold and Fay L. Faurote, *Ford Methods and Ford Shops* (New York: Engineering Magazine Co., 1915), pp. 41–42.

25. During the early Highland Park years, the company employed few women. Women worked at clerical jobs in the administration building, in the upholstery shop, and a few in small magneto assembly; none worked at the jobs described here. In general, Ford felt that "women had no place in the factory." See Kathleen Anderson Steeves, "Workers and the New Technology: The Ford Motor Company, Highland Park Plant, 1910–1916" (Ph.D. diss., George Washington University, 1986), p. 125.

26. See David Gartman, *Auto Slavery* (New Brunswick, N.J.: Rutgers University Press, 1986), for a discussion of discretion in auto work.

27. Fay L. Faurote, "Special Ford Machines," report for the Additional Tax Case, Acc. 96, box 7, Ford Archives. Though the figures are wrong, the passage gives the reader a clear idea of the principles of Fordism.

28. Ibid.

29. Martin La Fever, "Workers, Machinery, and Production in the Automobile Industry," *Monthly Labor Review* 19 (Oct. 1924): 735–60.

30. Meyer, *Five Dollar Day,* p. 80; Hildebrand, *Designing for Industry,* p. 2; Albert Kahn Papers, Detroit Institute of Arts; Engineering Appraisal, AR 65-71:22, HP 1949, Ford Industrial Archives, Redding, Mich.; "Movement Costs," *Automotive Topics* 40 (June 22, 1916): 1082.

31. Rockwood Conover, "The Factory Transportation of Production and Materials," *American Machinist,* June 15, 1916, p. 1036.

32. Oliver J. Abell, "Making the Ford Car," *Iron Age* 89 (June 6, 1912); 1384; note that Abell wrote a series of articles under this title for *Iron Age.*

33. Nevins, *Ford: The Times,* p. 383; Meyer, *Five Dollar Day,* p. 10; *Factory Facts from Ford,* 1917, Ford Archives.

34. Arnold and Faurote, *Ford Methods*, p. 25; Oscar Bornholt, "Placing Machines for Sequence of Use," *Iron Age* 92 (Dec. 4, 1913): 1276; Harry Jerome, *Mechanization in Industry* (New York: National Bureau of Economic Research, 1934), p. 188.

35. Oliver Abell, "Making the Ford Car," *Iron Age* 92 (Dec. 1913): 1276–77.

36. Meyer, *Five Dollar Day*, p. 48; also see Fisher Body Corp., Job Descriptions, Mary Van Kleek Collection, box 32, folder 9, Walter Reuther Library, Wayne State University, for general descriptions and wages of auto workers.

37. Arnold and Faurote, *Ford Methods*, p. 25.

38. Abell, "Making the Ford Car," *Iron Age* 89 (June 6, 1912): 1386.

39. Ibid., 92 (July 3, 1913): 2.

40. Arnold and Faurote, *Ford Methods*, p. 38.

41. "A. M. Wibel Reminiscences," Ford Archives; Robert Cummings, Address to Engineering Society, May 17, 1910, Cummings Papers, box 27 Division of Engineering and Industry, Smithsonian Institution; "Weismyer Reminiscences" and "Wibel Reminiscences," Ford Archives.

42. Wibel Interview, Feb. 11, 1913, Acc. 940, box 22, Ford Archives.

43. Bornholt, "Placing Machines," 1276–77.

44. Arnold and Faurote, *Ford Methods*, pp. 29–30; "William Pioch Reminiscences," Ford Archives.

45. "Pioch Reminiscences."

CHAPTER SIX. THE RATIONAL FACTORY: HIGHLAND PARK'S
NEW SHOP, 1914–1919

1. "Production by Years," Acc. 33, box 41, Ford Archives.

2. Figures based on 1909 and 1914 census, cited in *Iron Age* 97 (Feb. 24, 1916): 499.

3. "Operating Details of a Large Detroit Plant: Handling Material and Recording Labor in the Works for Making Automobile Parts Described Last Week," ibid. 91 (Jan. 9, 1913): 144; at the time, the Dodge brothers were still making parts exclusively for Ford and had not yet begun to build their own car.

4. Harry C. Spillman, "Factory Building Equipment Details: Partitions, Storage, Racks, Work Benches, and Other Points of Wide Application in Plant of Continental Motor Manufacturing Co., Detroit," ibid. (Mar. 6, 1913): 581; "Plant of the Eastern Car Co.," ibid. 94 (Sept. 24, 1914): 703.

5. William Knudsen, "Assembly Department Report," 1914–15, Acc. 1, box 122, Ford Archives.

6. Horace L. Arnold and Fay L. Faurote, *Ford Methods and Ford Shops* (New York: Engineering Magazine Co., 1915), p. 25; Henry Ford, *My Life and Work* (Garden City, N.Y.: Doubleday, Page, 1922), p. 39.

7. Arnold and Faurote, *Ford Methods*, p. 386.

8. L. V. Spencer, "Metamorphosis of the Motor Car," *Motor Age* 29 (Mar. 9, 1916): 5–11.

9. In 1916 two additional buildings, identical to the first two, were completed, giving Ford four parallel manufacturing buildings with craneways between each set of two.

10. Arnold and Faurote, *Ford Methods*, p. 25.

11. Memo, conference of Joseph Davis with Fay L. Faurote, June 1, 1926, Additional Tax Case, Acc. 96, box 12, Ford Archives.

12. Conference with P. Martin, Hartner, and Dagner, June 3, 1926, re Additional Tax Case, Acc. 96, box 12, Ford Archives; Henry F. Porter, "Four Big Lessons from Ford's Factory," *System* 31 (June 1917): 640–41; Davis, Faurote memo.

13. Oliver J. Abell, "Ford's Screw Machine," *Iron Age* 95 (Mar. 1915): 495; also see David Hounshell, *From the American System to Mass Production, 1800–1932: The Development of Manufacturing Technology in the United States* (Baltimore: Johns Hopkins University Press, 1984), pp. 230–33; Arnold and Faurote, *Ford Methods*.

14. Abell, "Ford's Screw Machine," p. 41.

15. Memo, interview with Wibel at Highland Park, July 23, 1926, Acc. 96, box 14, Ford Archives; "George Wollering Reminiscences," Ford Archives.

16. Philip Hanna, "Hudson Prospering through Efficiency," *Wall Street Journal*, Oct. 8, 1926.

17. *Ford Man*, Jan. 3, 1918, p. 1.

18. O. J. Abell, "The Making of Men, Motor Cars, and Profits," *Iron Age*, 95 (Jan. 1915): 37; "The Manufacturer Much to Be Admired," *Automobile Topics* 45 (Feb. 24, 1917): 254; Stephen Meyer, *The Five Dollar Day: Labor Management and Social Control in the Ford Motor Company 1908–1921* (Albany: State University of New York Press, 1981), p. 56.

19. "Wiesmyer Reminiscences," Ford Archives.

20. *Architectural Forum* 139 (1930): 90; *Ford Industries* (an in-house publication), 1924, p. 13; Henry T. Noyes, "Planning for a New Manufacturing Plant," *Annals of the American Academy of Political and Social Science* 85 (Sept. 1919): 87.

21. *Ford Industries*, 1924, 13.

22. Ibid.

23. O. J. Abell, "A New Development in Factory Buildings," *Iron Age* 93 (Apr. 9, 1914): 903.

24. All over Detroit, workers referred the Ford Motor Company as "Ford's" or "Mr. Ford's"; even in the 1980s that was the standard way to refer to the company.

25. Anonymous letter to *Ford Worker* (a union paper) 1 (1926): 3, quoted in Meyer, *Five Dollar Day*, pp. 40–41.

26. Oliver J. Abell, "Making the Ford Motor Car," *Iron Age* 89 (June 13, 1912): 1458.

27. Arnold and Faurote, *Ford Methods*, pp. 105–9.

28. Pat Greathouse, oral history taken by Jack Skeels, May 14, 1963, Walter Reuther Library, Wayne State University.

29. Greathouse oral history; Meyer, *Five Dollar Day*, p. 14.

30. Charles Reitell, "Machinery and Its Effects upon the Workers in the Auto Industry," *Annals of the American Academy of Political and Social Science* 116 (Nov. 1924): 37–43.

31. Meyer, *Five Dollar Day*, pp. 34, 51.

32. Reitell, "Machinery and Its Effects," p. 37; Meyer, *Five Dollar Day*, p. 42.

33. Meyer, *Five Dollar Day*, p. 56; William Lazonick, "Technological Change and Control of Work: The Development of Capital-Labour Relations in U.S. Mass Production Industries," in Howard F. Gospel and Craig R. Littler, eds., *Managerial Strategies and Industrial Relations* (London: Heinemann, 1983), p. 124; David Montgomery, *Workers' Control in America* (New York: Cambridge University Press, 1979), p. 102.

34. Hugh G. J. Aitken, *Scientific Management in Action: Taylorism at Watertown Arsenal, 1908–1915* (1960; reprint, Princeton: Princeton University Press, 1985), pp. 12, 232–34.

35. See Meyer, *Five Dollar Day*, pp. 87–89.

36. Joseph G. Rayback, *A History of American Labor* (New York: Macmillan, 1965), ch. 24; Philip S. Foner, *History of the Labor Movement in the United States* (New York: International Publishers, 1965), vol. 4.

37. Lazonick, "Technological Change," p. 124.

38. Samuel Haber, *Efficiency and Uplift* (Chicago: University of Chicago Press, 1964), p. 25; also see Meyer, *Five Dollar Day*.

39. Assembling Department Report, 1914–15, Aug. 11, 1915, Acc. 1, box 122, Ford Archives.

CHAPTER SEVEN. FORD'S MOST AMBITIOUS MACHINE:
THE RIVER ROUGE PLANT, 1919–1935

1. John Van Deventer, "Links in a Complete Industrial Chain," *Industrial Management* 64 (Sept. 1922): 131–32 (emphasis in original). Like the Highland Park plant, the Rouge had its chroniclers. In 1922 and 1923, *Industrial Management* magazine ran a ten-part series by John Van Deventer entitled "Ford Principles and Practices at River Rouge." He documented nearly every process and building at the plant. A decade later the editor of *Mill and Factory* wrote a less in-depth series on the plant.

2. Memo from John A. Moekle staff attorney, to F. J. Kallin, Plant Engineering Office, June 17, 1959, Files of Al Wowk, Ford Motor Company World Headquarters, Dearborn, Mich.

3. Charles Sorensen, *My Forty Years with Ford* (New York: W. W. Norton, 1956), p. 157.

4. Ford Motor Company Corporate Papers, Acc. 85, box 1, p. 246, Ford Archives.

5. William B. Mayo, "Report on Rouge River Location," Nov. 13, 1916, Acc. 62, box 49, Ford Archives.

6. Julian Kennedy to Ford Motor Company, Nov. 13, 1916, Acc. 62, box 49, Ford Archives.

7. Mayo, "Report."

8. Major H. Burgess, Corps of Engineers, "Preliminary Examination of Rouge River, Michigan," *Examination of Rivers and Harbors,* p. 15, U.S. Doc. 445, 64th Cong., 2d sess., 1916–17, H. Docs., vol. 22.

9. Colonel Frederick V. Abbot, Corps of Engineers, ibid., p. 4.

10. David Hounshell, "Ford Eagle Boats and Mass Production during World War I," in M. Roe Smith, ed., *Military Enterprise and Technological Change* (Cambridge: MIT Press, 1985), pp. 175–202.

11. Josephus Daniels, Secretary of the navy, to Henry Ford, Dec. 22, 1917, Senate Committee on Naval Affairs, re Eagle Boats, 66th Cong., 2d sess., 1919, S. 133-5, p. 6; "Supplemental Sheet," Acc. 572, box 26, Ford Archives.

12. David L. Lewis, *The Public Image of Henry Ford* (Detroit: Wayne State University Press, 1976), p. 100.

13. Ibid., pp. 99–102; Allan Nevins and Frank Hill, *Ford: Expansion and Challenge, 1915–1933,* vol. 2 of Nevins, *Ford* (New York: Scribner's, 1957), pp. 88–89. The company was involved in more than boat building on behalf of the war effort. The Highland Park plant produced trucks, tanks, and helmets for U.S. and Allied troops.

14. Moritz Kahn, "Plan the Plant for the Job," *Factory and Industrial Management* 75 (Feb. 1928): 316–18.

15. *Engineering News-Record* 81 (Oct. 17, 1918): 700; Sorensen, *Forty Years,* p. 170.

16. Sorensen, *Forty Years,* pp. 165–67; *Ford Man,* Sept. 3, 1919; "William Pioch Reminiscences," p. 23, Ford Archives.

17. Sam Bass Warner Jr., *Street Car Suburbs: The Process of Growth in Boston* (Cambridge: Harvard University Press, 1962).

18. Undated company memo, Files of Al Wowk, Ford Motor Company World Headquarters, Dearborn, Mich.

19. Michael Mahoney, "Reading a Machine: The Products of Technology As Texts for Humanistic Study" (manuscript), p. 14.

20. *Ford Man,* Oct. 3, 1919.

21. Sorensen, *Forty Years,* pp. 172–74; Lewis, *Public Image,* ch. 10.

22. Lewis, *Public Image,* ch. 10.

23. Ibid., pp. 163–64.

24. *Ford Industries,* 1924, p. 9.

25. *Iron Age* 102 (Dec. 10, 1918): 1520.

26. The loss of the plant engineering department records in the Rotunda fire in 1960s makes it impossible to explain the separation with certainty.

27. Roy S. Mascon, *Should the Office and Factory Be Separated,* Office Executive Series, no. 37 (New York: American Management Association, 1928); Grant Hildebrand, *Designing for Industry: The Architecture of Albert Kahn* (Cambridge: MIT Press, 1974); Hartley W. Barclay, *Ford Production Methods* (New York: Harper Bros., 1936).

28. E. G. Liebold to D. Boyer, July 25, 1922, Acc. 572, box 23, Ford Archives.

29. Barclay, *Ford Production Methods,* p. 99; John Van Deventer, "Mechanical Handling of Coal and Coke," *Industrial Management* 65 (May 1923), 196; John Van Deventer, "Machine Tool Arrangement and Parts Transmission," *Industrial Management* 65 (May 1923): 259.

30. Van Deventer, "Mechanical Handling," pp. 197–98.

31. Unsigned letter to Charles Sorensen, Aug. 2, 1929, Acc. 572, box 23, Ford Archives.

32. Barclay, *Ford Production Methods,* p. 100; Van Deventer, "Mechanical Handling," pp. 197–98.

33. Van Deventer, "Machine Tool Arrangement," p. 259.

34. "Pioch Reminiscences," p. 44.

35. Van Deventer, "Links," p. 133; Barclay, *Ford Production Methods,* p. 95.

36. The Detroit, Toledo, and Ironton was a regional freight line that Ford bought to supply the Rouge.

37. Barclay, *Ford Production Methods,* p. 99; Mary Jane Jacobs, "The Rouge in 1927," in *The Rouge* (Detroit: Detroit Institute of Arts, 1978), p. 23.

38. *Ford Industries,* 1924, p. 13.

39. Alfred D. Chandler, *Giant Enterprise* (New York: Harcourt Brace & World, 1964), p. 13.

40. *Ford News,* Oct. 1, 1927, p. 8; Oct. 14, 1927, p. 4.

41. Philip Scranton, *Proprietary Capitalism: The Textile Manufacture at Philadelphia, 1800–1885* (Cambridge: Cambridge University Press, 1983), ch. 4.

CONCLUSION

1. Secretary, State Board of Charities, to J. W. McLane, Sloane Maternity Hospital, Apr. 13, 1900, Presbyterian Hospital of New York Archives; as quoted in Charles Rosenberg, *The Care of Strangers: The Rise of America's Hospital System* (New York: Basic Books, 1987), p. 328.

2. I thank Carol Ann Vaughn and Keith Wailoo for their help with this section on medical history. Susan Reverby, *Ordered to Care: The Dilemma of American Nursing, 1850–1945* (Cambridge: Cambridge University Press, 1987); Sylvia Hoffert, *Private Matters: American Attitudes toward Childbearing and*

Infant Nurture in the Urban North, 1800–1860 (Urbana: University of Illinois Press, 1989); Richard Wertz and Dorothy Wertz, *Lying-In: A History of Childbirth in America* (New Haven: Yale University Press, 1989); Rosemary Stevens, *In Sickness and in Wealth: American Hospitals in the Twentieth Century* (New York: Basic Books, 1989).

3. Thomas P. Hughes, *American Genesis: A Century of Invention and Technological Enthusiasm* (New York: Viking, 1989), ch. 7; Reyner Banham *A Concrete Atlantis: U.S. Industrial Building and European Modern Architecture, 1900–1925* (Cambridge: MIT Press, 1986), ch. 3.

4. Hughes, *American Genesis*, pp. 317–22.

5. I thank Thomas O'Shea for his help on Joyce and *Ulysses*.

6. Frank Budgen, *James Joyce and the Making of "Ulysses"* (New York: Harrison Smith & Robert Haas, 1934), pp. 67–68.

7. James Joyce, beginning of the "Aeolus" episode ("In the Heart of the Hibernian Metropolis"), ed. Hans Walter Gabler (New York: Vintage Books, 1986), p. 96.

8. Joyce to Grant Richards, May 5, 1906, *Letters of James Joyce*, ed. Richard Ellman (New York: Viking Press, 1966), 2:134; conversation with Arthur Power, in *Conversations with James Joyce*, ed. Clive Hart (London: Millington, 1974), p. 98, cited in Derek Attridge, *The Cambridge Companion to James Joyce* (Cambridge: Cambridge University Press, 1990), p. 261; Richard Ellman, *James Joyce* (New York: Oxford University Press, 1983), p. 702.

9. William Carlos Williams, *The Wedge*, 1944, cited in Cicelia Tichi, *Shifting Gears: Technology, Literature, and Culture in Modernist America* (Chapel Hill: University of North Carolina Press, 1987), p. 267.

10. "French Artists Spur On an American Art," *New York Tribune*, Oct. 24, 1915, quoted in Karen Lucic, *Charles Sheeler and the Cult of the Machine* (Cambridge: Harvard University Press, 1991), p. 29.

11. Lucic Sheeler, pp. 90–92.

12. Leo Marx, *The Machine in the Garden: Technology and the Pastoral Idea in America* (New York: Oxford University Press, 1964), pp. 355–56.

13. John Van Deventer, "Links in a Complete Industrial Chain," *Industrial Management* 64 (Sept. 1922): 137.

14. "By Their Works Ye Shall Know Them," *Vanity Fair*, Feb. 1928, p. 62; Van Deventer, "Links," p. 131.

15. Papers of Charles Sheeler, NSh 1, frame 101, Archives of American Art, Smithsonian Institution, cited in Mary Jane Jacobs, "The Rouge in 1927," in *The Rouge* (Detroit: Detroit Institute of Arts, 1978), p. 11.

16. Bertram Wolfe, *Diego Rivera, His Life and Times* (New York: Knopf, 1939), p. 313.

17. Linda Downs, "The Rouge in 1932," in *The Rouge*, p. 47.

Note on Sources

PRIMARY SOURCES CONSULTED FOR THIS BOOK INCLUDE EARLY BOOKS, NEWS-
papers, and journals, as well as manuscript collections. In thinking about the
beginnings of the idea of the rational factory and its development during the
nineteenth century, I have found Andrew Ure, *Philosophy of Manufactures or An
Exposition of the Scientific, Moral, and Commercial Economy of the Factory System
of Great Britain* (London, 1835); Charles Babbage, *On the Economy of Machines
and Manufactures* (London: Charles Knight, 1835); Oliver Evans, *The Young Mill-
Wright and Miller's Guide* (Philadelphia, 1795); and *Niles' Weekly Register* (Baltimore and Washington, D.C.), especially helpful.

Information on industrial engineers' education and work came from the MIT
Archives and a number of engineering magazines: *Iron Age* (New York); *American Machinist* (New York); *Engineering Magazine* (New York); *Technology Review*
(Cambridge, Mass.); and *Proceedings of the Society of Industrial Engineers* (New
York). During the early years of the twentieth century, industrial engineers wrote
dozens of books describing their work (see notes in chapter two). These sources
prove important in learning about engineers' goals, both stated and unstated, in
redesigning factories.

In writing about the automobile industry, the Ford Motor Company, and the
builders of its factories, I have relied most heavily on the manuscript collections
of the Ford Motor Company housed at the Edison Institute in Greenfield Village,
Dearborn, Michigan. The Detroit Institute of Arts holds some papers of Albert
Kahn which are essential to any discussion of factory architecture in the United

States. Horace Arnold and Fay Faurote, *Ford Methods and Ford Shops* (New York: Engineering Magazine Co., 1915), is an essential beginning point for anyone interested in early Ford operations. *Industrial Management* (New York) and *Mill and Factory* (New York), along with the engineering magazines cited above, were also important in putting together the story of changes in the auto industry.

Any study of the rationalization of American industry must pay tribute to Sigfried Giedeon and Lewis Mumford, whose seminal works on mechanization and architectural design deeply inform this book. See Giedeon, *Mechanization Takes Command: A Contribution to Anonymous History* (New York: Oxford University Press, 1948); Mumford, *Technics and Civilization* (New York: Harcourt, Brace & World, 1963); Mumford, *The Myth of the Machine: The Pentagon of Power* (1964; reprint, New York: Harcourt Brace Jovanovich, 1970). The literature on factory architecture itself is small. important works include: Reyner Banham, *A Concrete Atlantis: U.S. Industrial Building and European Modern Architecture* (Cambridge: MIT Press, 1986); Grant Hildebrand, *Designing for Industry: The Architecture of Albert Kahn* (Cambridge: MIT Press, 1974). The literature on the technology, work, and history of the auto industry, on the other hand, is so voluminous that one can mention only a few titles, any of which will lead the interested reader farther: Allan Nevins's three-volume work *Ford* (New York: Scribners, 1954–63) is the classic work on the Ford Motor Company; David Hounshell, *From the American System to Mass Production, 1800–1932: The Development of Manufacturing Technology in the United States* (Baltimore: Johns Hopkins University Press, 1984), provides the most thorough treatment of the development of the technology and managerial strategies that came together in the auto industry; David Montgomery, *Workers' Control in America: Studies in the History of Work, Technology, and Labor Struggles* (New York: Cambridge University Press, 1979), usefully explores important issues inside the factory. Stephen Meyer, *The Five Dollar Day: Labor Management and Social Control in the Ford Motor Company, 1908–1921* (Albany: State University of New York Press, 1981), provides a detailed look at work in the Ford factories.

Index

Page numbers in italic denote illustrations; those in bold face denote tables.